T0334591

Primer on Enhanced Oil Recovery

Primer on Enhanced Oil Recovery

Vladimir Vishnyakov
Institute for Materials Research, University of Huddersfield,
Huddersfield, United Kingdom

Baghir Suleimanov
Oil and Gas Scientific Research Project Institute,
State Oil Company of Azerbaijan Republic (SOCAR),
Baku, Azerbaijan

Ahmad Salmanov
Oil and Gas Scientific Research Project Institute,
State Oil Company of Azerbaijan Republic (SOCAR),
Baku, Azerbaijan

Eldar Zeynalov
Institute of Catalysis and Inorganic Chemistry,
Azerbaijan National Academy of Sciences (ANAS),
Baku, Azerbaijan

Gulf Professional Publishing
An imprint of Elsevier

Gulf Professional Publishing is an imprint of Elsevier
50 Hampshire Street, 5th Floor, Cambridge, MA 02139, United States
The Boulevard, Langford Lane, Kidlington, Oxford, OX5 1GB, United Kingdom

Notices
Knowledge and best practice in this field are constantly changing. As new research and experience broaden our understanding, changes in research methods, professional practices, or medical treatment may become necessary.

Practitioners and researchers must always rely on their own experience and knowledge in evaluating and using any information, methods, compounds, or experiments described herein. In using such information or methods they should be mindful of their own safety and the safety of others, including parties for whom they have a professional responsibility.

To the fullest extent of the law, neither the Publisher nor the authors, contributors, or editors, assume any liability for any injury and/or damage to persons or property as a matter of products liability, negligence or otherwise, or from any use or operation of any methods, products, instructions, or ideas contained in the material herein.

British Library Cataloguing-in-Publication Data
A catalogue record for this book is available from the British Library

Library of Congress Cataloging-in-Publication Data
A catalog record for this book is available from the Library of Congress

ISBN: 978-0-12-817632-0

For Information on all Gulf Professional Publishing publications
visit our website at https://www.elsevier.com/books-and-journals

Publisher: Joe Hayton
Senior Acquisitions Editor: Katie Hammon
Editorial Project Manager: Aleksandra Packowska
Production Project Manager: Selvaraj Raviraj
Cover Designer: Christian J. Bilbow

Typeset by MPS Limited, Chennai, India

Working together
to grow libraries in
developing countries

www.elsevier.com • www.bookaid.org

Contents

About the authors

Vladimir Vishnyakov

Vladimir Vishnyakov PhD, SPE, FInstP. He has an extensive track record in physics, materials research and surface analysis. He lectures and supervises postgraduate students in enhanced oil recovery methods and techniques at the University of Huddersfield.

Affiliation and expertise

Professor, Director of Institute for Materials Research, University of Huddersfield, Huddersfield, United Kingdom

Baghir Suleimanov

Bagir Suleimanov DSc. He is an expert in oil field development and oil recovery processes. He is Deputy Director for oil and gas production at the Oil and Gas Scientific Research Project Institute within the State Oil Company of Azerbaijan Republic (SOCAR), Baku, Azerbaijan.

Affiliation and expertise

Professor, Corresponding member of National Academy of Sciences (ANAS), Baku, Azerbaijan

Ahmad Salmanov

Ahmad Salmanov DSc. He has comprehensive expertise in paleogeology and oil reservoir modelling. He is Deputy Director for geology and reservoir development at the Oil and Gas Scientific Research Project Institute within the State Oil Company of Azerbaijan Republic (SOCAR), Baku, Azerbaijan.

Affiliation and expertise

Deputy Director, Oil and Gas Scientific Research Project Instiute, Baku, Azerbaijan

Eldar Zeynalov

Eldar Zeynalov DSc. He has been working for many years in the area of oil chemistry and mechanisms of oil−nanoparticle interactions. He is leading the Carbon Nano Catalysis Laboratory at the Institute of Catalysis & Inorganic Chemistry ANAS, Baku, Azerbaijan.

Affiliation and expertise

Professor, Corresponding member of National Academy of Sciences (ANAS), Baku, Azerbaijan

Introduction

1

Abstract

Crude oil is a vital natural resource for civilization. Traditional oil extraction is a complex process and on the industrial scale complete extraction of oil from a reservoir is impossible. Less than half of reservoir oil, even in the best case scenario of light oil, can be extracted "relatively" easily. High viscosity crudes are almost unextractable by the traditional methods. Enhanced Oil Recovery (EOR) is a set of technologies and methods to extract more oil and progress extraction up to 80% and beyond. This book is an introduction to EOR. It covers all basic processes and concepts at the beginning and then introduces all EOR traditional and newly used techniques, including a brief on renewable energy utilisation in EOR. The presented material is very condensed and full subject understanding is only possible on the basis of many other texts, original publications and practical experience.

Crude oil extraction is complicated and is stretched-in-time business. Big profits and losses can be easily made. Good understanding of processes and careful, both technological and business, planning are essential. The book only marks the boundaries of this highly professional field.

The book does not require any prior knowledge in oil extraction and is intended for a wide audience.

Oil was, is and will be for long time an essential natural resource for the human civilization. Oil has been and is used as an energy source and as a chemical commodity. It is estimated that currently every day we consume almost 100 thousand barrels of oil. It is said that burning oil and all it derivatives supplies approximate a third of all energy used by the humanity. While our estimates for oil reserves vary wildly and depend on accounting for different types of oil and different extraction methods, two points remind the same. Oil is not a renewable resource — eventually we will use majority of it and we urgently need to seek other energy sources for our activities. Secondly, burning oil and other hydrocarbons releases very high amounts of carbon dioxide into the atmosphere. Carbon dioxide capture, while developing rapidly, most likely would not capture all produced carbon dioxide and this greenhouse gas will accumulate in the atmosphere with all disastrous consequences of global warming. On both accounts, it will be prudent to limit and reduce oil use for energy generation. However, the forecast in oil consumption rather predicts an increase in global oil usage.

Oil is produced by extraction from predominantly underground reserves. We more and more extinguish easily reachable reservoirs and we need to involve more and more advanced technology to get oil for the refineries and other chemical plants. Market oil price is a very complex composite and is dictated by many

Primer on Enhanced Oil Recovery. DOI: https://doi.org/10.1016/B978-0-12-817632-0.00001-3

natural, market and political forces. However, what is inevitable is the rise of oil production price. This is dictated by more difficult extraction conditions, growing complexities of extraction technologies and the need for an oil fields infrastructure maintenance, upgrades and developments.

One of the most important facts about oil extraction is that we newer can extract all oil from the real oil field, e.g. from the underground reserve. Depending on our luck and with not so great difficulties we can extract only 10−30% of the oil in place. With a bit of well-developed technology and additional efforts we can push our extraction to around 50%. With all our technology on the industrial scale we would most probable not be able to get much above 80% of extracted of oil in place.

The last 25−30% of oil in place extraction (remember that almost 20% of oil we would never be extracted) needs employment of all knowledge and skill we have developed and accumulated. The additional deployment and production technologies most likely can cost us up to $20 per barrel. It is immediately clear that the technology use can significantly reduce profit margins. Complex economic modeling and forecasting need to be used to sustain profitability in the oil production process. Political and local considerations also come into the account. All this makes oil extraction a very complex business indeed.

This book is about our desire to extract above-mentioned additional 25−30% of oil in place after straightforward techniques extract first 50% of oil. Oil extraction becomes quite complex at this stage so it is named Enhanced Oil Extraction (EOR). The authors of this book tried their best to guide the reader to build his/her knowledge from simple concepts to the basic understanding of EOR techniques. There are many books and texts on EOR methods including classic book by D.W. Green and G.P. Wilhite. So why another book? We, the authors, have tried our best to reflect on the current pool of people involved in oil production at all stages and from all business sides. More and more people without broad technical knowledge such us accountants, bankers, environmentalists, politicians and public, in some cases, are involved in policies forming and decision taking. We in any measure do not judge their technical knowledge but we tried our best to widen their knowledge horizon.

We, the authors, all have done our best to present material on the introductory but appropriate level. VV had initiated this book writing, generally planned it and overlooked the process. VV has also predominantly written Chapters 1, 3, 4, 5, 9, 14 and 17. BS has primarily written chapters 7, 8, 10, 11, 12, 13, 15 and 16. AS has principally written chapters 2 and 6. EZ took part in writing Chapters 3 and 12.

Having great respect for mathematics we nevertheless used it in the book as little as possible. This has been done on the reflection that many things can be reasonably well known without familiarity of exact mathematical ways to arrive to certain conclusions. At the same time, the average complexity equations do not help much to arrive to technically applicable results. Detailed description, on the other side, is very complex and should to be left to the people who do it every day. We are sure that if the readers would get a knowledge presented in this book and then decide to deepen their understanding of the physical or chemical phenomena, they will have no big trouble to find waste number of original texts and papers to guide and help them. Most likely, the best information sources are professional societies and the reader would be strongly advised to join one.

It is in human general nature to remember well-structured facts; we usually refer to this as logically bonded knowledge. For this reason, we have written this book starting from simple phenomena and processes. Little by little, the phenomena will get more complicated but we hope that each step is well justified and is easily understood.

We have to stress again that in reality the Enhanced Oil Extraction and all used in it techniques invoke very complex natural and technological processes when applied to the real oil fields. Nevertheless, not understanding everything newer stopped human beings to get involved and to apply limited knowledge with a reasonably good rate of success. Oil recovery is complex and multistage process. The initial investment and price of mistakes are huge. We hope that by reading this book, the esteemed reader will make a first step in appreciation and comprehension of oil extraction complexity and, in some ways, this step would be just enough to make an informed, better judgement and this will pave the way to successful business.

Oil recovery and especially Enhanced Oil Recovery are energy-hungry processes. Energy use traditionally linked to CO_2 emissions. Many routes in the Enhanced Recovery rely on chemical processes and require injection of chemical agents into an oil formation. Environmental impact of all this is not negligible. Development of renewable energy utilization and biodegradable chemicals are but the responsible approach to the future of the industry.

Hydrocarbon and oil reserves classification

2

Abstract

Oil, gas and condensate are natural mixtures of hydrocarbon and non-hydrocarbon blends. Few expressions such as oil, crude oil, crude and petroleum are used interchangeably. The entomology of the last one has its roots in Greek. It is produced from "petro", which means "stone", and "oleum" — "oil". It is used to mean crude oil. In everyday life oil most usually refers to any viscous liquid.

2.1 Definition of hydrocarbon products

Oil, gas and condensate are natural mixtures of hydrocarbon and non-hydrocarbon blends. Few expressions such as oil, crude oil, crude and petroleum are used interchangeably. The entomology of the last one has its roots in Greek. It is produced from "petro", which means "stone", and "oleum" — "oil". It is used to mean crude oil. In everyday life oil most usually refers to any viscous liquid.

Crude oil is a natural oily (viscous) flammable liquid with a specific odor. It contains mainly hydrocarbons such as alkanes (linear molecules, single bonds,

Primer on Enhanced Oil Recovery. DOI: https://doi.org/10.1016/B978-0-12-817632-0.00002-5

Table 2.1 Elemental composition of crude oil and natural gas.

Element	Crude oil (at%)	Natural gas (at%)
Carbon	82–88	66–79
Hydrogen	11–15	1–25
Sulfur	0.05–2	0–0.18
Nitrogen	0.06–2	1–16
Oxygen	0.2–2	0

C_nH_{2n+2}, known also as paraffines), alkenes (linear molecules, double bonds, C_nH_2, commonly known as olefins) and arenes (molecules with aromatic ring(s), commonly known as naphthenes or cycloparafines). Properties of oil (viscosity, for instance) depend on the chemical composition, temperature and pressure on the first place.

The color of crude oils varies in broad spectra from yellow to dark brown, almost black, but there is also oil that has a yellow-green, brown-red color. In some formations oil is even colorless. The color and smell of oil is mostly defined by presence in the oil natural mixture of nitrogen-, sulfur- and oxygen-containing compounds. Some ideas about oil and gas elemental composition can be formed on the basis if Table 2.1.

Many properties of oil related to its density (weight per volume). Not all oil properties can be easily bunched together under this banner, nevertheless, most commonly, oil density can be used as a first step guidance, it represents marketable properties and, in some aspects, oil monetary value. Historically and practicable the oil density is expressed in relation to the density of fresh water at the standard temperature (see later).

Most commonly used scale expresses oil density in so named API (American Petroleum Institute) scale,

$$\text{API gravity } [^0] = (141.5/\rho) - 131.5, \tag{2.1}$$

where ρ is specific gravity, also known as density. All measurements should be done at the standard temperature $-$ 60°F (15.6 °C). These measuring conditions are known as "standard conditions". Gravity of fresh water on this scale is 10 ^{0}API (Fig. 2.1).

It is possible easily to use oil gravity to account for barrels of crude (volumetrically 1 barrel of oil crude contains 159 L of oil) in metric tonne.

$$N_{b/t} = (\text{API} + 131.5)/141.5 * 0.159 \tag{2.2}$$

Oil volume as a rule are measured in barrels. One barrel volume is equal to 42 USA gallons volume at the standard conditions.

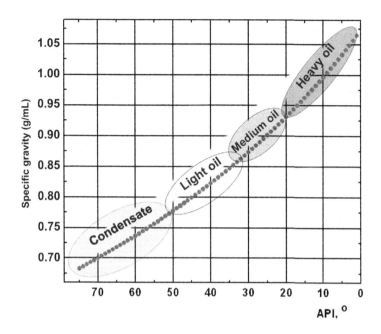

Figure 2.1 Link between API and specific gravity (density).

For example, one metric tonne of West Texas Intermediate (density 39.6 °API) contains at around 7.6 barrels in metric tonne. For oil known as Azeri light (density 34 °API) the barrel count would be at around 7.4.

When oil is in the oil containing formation it can have in it some dissolved gases and some chemical compounds containing many elements – sulfur, nitrogen and oxygen are most common. Sulfur is one of the most important and mostly unwanted elements in oil. From this point of view oil are divided on: low sulfur containing (up to 0.5%), medium sulfur containing (0.5–2%) and high sulfur oil (above 2% of sulfur). In many cases this oil classification is further simplified and oil is divided to "sweet" (below 0.5% sulfur) and "sour" (above 0.5%). Sulfur in oil complicates oil extraction and further rectification. Taking this into account, "sweet" oils can command price premium up to approximately 20%.

Oils are also divided by type of prevailing hydrocarbon types. There are various classifications. Some classifications contain three, some contain four members. For instance, in four member set there are four significant groups – paraffin-type oils, naphthene-type oils, asphalt-type oils and mixed-base oils.

Apart from carbon arrangements in the molecules, the hydrocarbons properties (oil properties) are affected by the molecular weight. The molecular weight on first place is defined by the number of carbons in the molecule. Small molecular weight compounds are gaseous at room temperature and pressure, medium weight compounds are liquids, while heavy compounds are solids. Light compounds produce more heat during burning (as relative percentage of hydrogen is higher) and for this reason command higher price when sold by the weight.

Concentration of some metals in oil (such as vanadium, titanium, nickel and so on) in some heavy (bitumen) oil formations is so high that it merits industrial extraction.

In the formation an oil is under higher than atmospheric pressure. The temperature is also usually higher than the room temperature (e.g. above 20 °C). High pressure allows oil to contain higher quantities of dissolved gas. All this leads to the situation that oil properties, in substantial part affecting the extraction, in the formation are significantly different compared to the same liquid after extraction and primary separation from water and gases.

It is clear at this point that an oil has many properties, both physical and chemical. One can count this as: chemical composition, molecular weight, density, viscosity, solidification and boiling points (temperatures), dissolved gas content, saturated gas pressure, coefficients of thermal expansion, volumetric coefficient and so on.

Oil volume as a rule is measured in barrels. One barrel volume is equal to 42 USA gallons volume at 60 °F (15.5 °C).

2.1.1 Natural gas

Natural gas is a mixture of hydrocarbon and other chemical molecules. Natural gas in the formation can be in the form of free gas (gas cap) and/or it can be dissolved in oil and in connate water. Natural gas at high pressure and a relatively low temperature can be in the formation even in crystalline form (gas hydrates). At room temperature and pressure natural gas is always in gaseous phase.

Generally natural gas contains methane (CH_4) at concentrations in 70−98%, ethane (C_2H_6), propane (C_3H_8) and butane (C_4H_{10}). There is also admixture of hydrogen sulfide (H_2S), carbon dioxide (CO_2), nitrogen (N_2) and helium (He). Some other inert gases also can be presented as well as even mercury in some cases.

As with oil, the main physico-chemical gas parameters are numerous. On the first place is, possibly, the molecular weight, as this inevitably leads to the amount of produced heat. Then there are: density at atmospheric pressure, average critical temperature and pressure, viscosity and so on.

Cleaned of impurities natural gas is odorless and colorless. To help in detection of gas leaks some substances with unpleasant smell − odorants − are added into natural gas before distribution to the end users.

The natural gas volume is usually measured in cubic feet (Ccf) in standard conditions (60 °F and 14.7 psi). Sometimes this changed to a measurement in metric units − cubic meters at room temperature (20 °C or 68 °F).

The measuring customer units in many cases depend on the sales region or the country. In the USA the sales are done in cubic feet (Ccf), in the Great Britain in kilowatt-hours, in Russia in cubic meters.

2.1.2 Condensate

Condensate is a mixture of light hydrocarbons with relatively low molecular weight which exists as a liquid at standard conditions (standard temperature and pressure). Chemically condensate does contain light carbohydrates but not propane and ethane. In some cases condensate contains small amounts of sulfur compounds.

Condensate low density leads to high API gravity numbers, which are in the region of $50-120\ ^0$API. Condensate comes from an oil well as so called associate gas (wet gas) and is also refereed at certain stages as lease condensate. In an oil formation condensate can exist as dissolved in oil/connate water gas or as a separate phase. All condensate behavior in the formation is fully defined by the formation chemistry, temperature and pressure. If the formation pressure during an oil production is not maintained by water or gas injection and formation pressure is lowered then condensate can start consolidating as a separate phase.

One of the main parameters to characterise condensate are condensate-gas factor and the pressure of the condensation onset.

The same as oil, condensate volume is measured in barrels at the standard conditions.

2.2 Oil reserve classification

Development of human society at large extend depends, at the present, on availability of non-renewable hydrocarbons (gas, oil and bitumen). Many factors and technologies need to be taken into consideration for hydrocarbon extraction. The oil production includes complex interplays of geology, technology, social, financial, political and environmental challenges. All this needs to be taken into the consideration in oil exploration, extraction and processing.

In order to evaluate commercial viability of hydrocarbon formation, on the first place, a complex geological study need to be undertaken. Various international and national classifications and schemes can be applied to indicate industrial significance of proven hydrocarbon reserves in the formation.

Existing classifications are vigorously set and are closely monitored by the states (governments), fiscal organs, banks and companies management. All this monitoring allows for a well-structured set of geological and exploitation steps. Reserve management then forms the basis of the reserve mineralogical and energy potential. The potential and development strategies are crucial for the reserve owners and developers.

The classification of hydrocarbon reserves (resources) (gas, oil, bitumen) establishes uniform principles for calculating and assessing readiness for industrial development. Classification is an extremely important undertaking and much consideration and attention are given to it. Along with the generally accepted international classifications such as SPE/WPC/AAPG/SPEE ("SPE-PRMS") and SEC, which are used uniformly by all the oil companies of the world, other classifications exist. In almost every oil-producing country there are also national classifications (SORP, CSA, PRO, NPD, USGS, to mention just few).

In the 1990s European Economic Commission (EEC) began work to develop and set universal standards to classify solid fuels and mineral reserves. Since 2004 the classification has been extended to hydrocarbons (natural gas and oil). At the same time the classification was internationally adopted and renamed. Now it is known as the United Nations Framework Classification of Fuels and Minerals Stocks (UNFC).

The calculated volumes of hydrocarbons in the earth crust are regarded as "reserves-resources". The volumes are presented in open and unopened natural deposits (recoverable and non-extractable resources and reserves), plus already extracted hydrocarbons. Quantitative estimation of "resource reserves" is made by calculating the total volume of hydrocarbons. Both, the identified deposits, with geological proven oil and gas content, and in promising deposits, that are expected to open in the future, are accounted for. The term "reserves-resources" refers to all types of hydrocarbon raw materials that are currently referred to as "traditional" or "unconventional" reserves. "Unconventional" refers to deposits of tight gas and oil in solid sandstone, shale oil and gas, gas hydrate accumulations, and so on.

The assessment of reserves aim to get oil volume in barrels and gas reserve in cubic feet or cubic meters. This is potential amount for commercial exploration. Depending on the classification schemes and perceived uncertainties (risks) of future production, hydrocarbons in the formation (oil fields) are grouped into different categories, for instance: petroleum reserves; contingent resources and prospective resources (see Fig. 2.2).

One of the most widely used schemes for classification and management of hydrocarbon reserves in the world is SPE-PRMS. This is mostly due to the fact that the scheme takes into account the greatest number of variables (factors) during the forecasted production. Also, the scheme makes the distinction between the main categories of recoverable hydrocarbon reserves and resources, which include: "Production", "Reserves", "Conditional resources" and "Prospective resources" as well as "Unrecoverable" amounts of hydrocarbons.

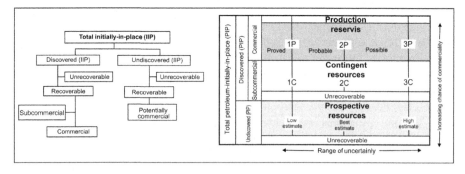

Figure 2.2 Hydrocarbon reserve classification.

In the following the main classes and categories of geological hydrocarbon reserve volumes are listed.

The total initial geological resources and reserves are the initial volumes of hydrocarbons found in natural deposits before the start of their development, as well as the volumes of hydrocarbons in deposits that are expected to discovered (forecasting) in the future, the equivalent of "total resources".

Open initial geological resources and reserves are calculated hydrocarbon volumes contained in the already known (explored) deposits prior to the beginning of commercial exploration of hydrocarbons.

Production – accumulated volumes of liquid, gaseous and solid hydrocarbons, produced on a specific date, e.g. already extracted hydrocarbons.

Reserves – are the amount of oil, combustible gases, natural condensate and associated components contained in them, which is located in the deposits studied by drilling and which are supposed to be mined from a specified date under existing economic conditions. All estimates of reserves take into account a certain degree of risk or uncertainty, which depends on the number of geological, geophysical and field data, technical and economic, available. All factors should be accounted for at the time of calculating reserves. This will include all necessary financial data: cost of area exploration, the cost of drilling, transportation costs, set prices per barrel, taxes, and much more. All this plays a role in determining the feasibility of a project to develop reserves.

Conditional resources are volumes of potentially recoverable hydrocarbons, calculated from the assessment (as of a specific date) of explored deposits, but with delayed production. The delay is due to existing extraction constraints. Conditional resources are classified according to the level of confidence determined during the modeling. Modeling also takes into account the phases of the project and it economic indicators. If restrictive conditions can change in the future and conditional resources can be profitably developed, then these hydrocarbon volumes should be transferred to the other appropriate categories.

For example, conditional resources may include projects for which, at present, for examples: there are no profitable markets, hydrocarbon production depends on the emergence of new technologies, or on which the calculation of reserves (resources) of a deposit does not allow an unequivocal conclusion about the industrial significance of a development project.

Undiscovered initial oil-in-place – the amount of liquid, gaseous and solid hydrocarbons contained in deposits that are not yet discovered (but forecasted as possible) on the date of calculation.

Perspective resources – are the volumes of potentially recoverable hydrocarbons which are calculated when assessing (on a specific date) exploration. These are prospective structures, i.e. undiscovered deposits for the future implementation of future development projects.

For example, prospective resources may be in the deep-lying horizons of the developed fields but not discovered at the date of the estimate. Nevertheless, there might be favorable geological conditions for the accumulation of hydrocarbons.

An important element in calculating prospective resources is the determination of risk or uncertainty, which must be calculated by probability theory.

Prospective resources are accessed by two probabilities - the likelihood of their discovery and the likelihood of the development.

Prospective resources are classified according to the level of reliability determined during the assessment of recoverable resources. They are subject to their finding and later industrial production, as well as good fitment into the stages of the project development.

Unrecoverable resources – are the part of the discovered and unexploited initial geological reserves and hydrocarbon resources (discovered or undiscovered initial-oil-in-place). The resources are accounted according to an estimate (as of a specific date). The resources cannot be extracted in future development projects of the formation.

However, a part of the specified amount of unrecoverable resources may become recoverable in the future as the economic situation changes or new extraction technologies are developed. Whatever left might remain in the reservoir and can be not recoverable due to geological, physical and chemical constraints.

Estimated Ultimate Recovery (EUR) in the classification SPE-PRMS strict interpretation is not a category of reserves and resources. However, this term can be used to estimate (on a specific date) the amount of hydrocarbons potentially recoverable (extractable) from any exploited or untouched reservoir.

When conducting studies to assess the amount of resources in an oil containing location and to denote total geological reserves and resources, alternative terms such as **Total Resource Base** or **Hydrocarbon Endowment** can be used. Total recoverable reserves or NIH may be referred to by the term "area potential". Total recoverable or EUR may be termed Basin Potential.

The sum of reserves, conditional resources and prospective resources, can also be defined as the term "remaining recoverable resources".

When using these terms, it is necessary to ensure that each of the components of the Summing Stocks and Resources are consistent with the various degrees of technical and economic risk given the definitions in the SPE-PRMS classification.

The SPE-PRMS classification divides stocks into categories: "possible", "probable" and "proved."

Each category in practice lead to average successful extraction rate, as proven by many already implemented projects and strict assessment criteria. "Proved" category is extracted in 90% of cases. "Probable" reserves are extracted with 50% susses. Only 1 in 10 cases is successful in extraction of "possible" category.

Proved reserves (P90) — reserves that can be extracted under existing economic conditions and applied technological processes, using probabilistic models, there should be a 90% probability that the cumulative production will be equal to or exceed the estimated reserves.

The proven reserves developed by Proved Developed Producing (PDP) will be extracted from the developed reservoirs that have been discovered and are being developed at the time of their assessment.

Assessing uncertanty

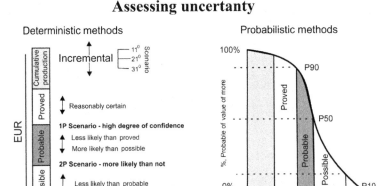

Figure 2.3 Reserves uncertainty classification.

Proved reserves that are not developed [e.g. Proved Developed Nonproducing (PDN)] consist of reserves confined to unopened (not drilled yet) formation zones and to non-perforated wells.

Proved Undeveloped (PUD) reserves will require drilling of new wells and significant capital expenditures are expected.

Probable reserves (P50) are reserves that, according to the geological data, are potentially more recoverable than not. In this context, when using probabilistic methods, there should be at least a 50% chance that the actual amount of oil recovered will be equal to or exceed the amount of proved and probable reserves.

Possible reserves (P10) are unproved reserves, the ability to extract which is less than probabilistic reserves. When a probabilistic estimation method is used, there must be at least a 10% chance that the actual amount of oil produced will be equal to or exceed the sum of proven, probable and possible reserves.

On Fig. 2.2 the reserve classification is joined with range of uncertainty. The uncertainty reflects the intervals of change of the calculated quantities (volumes) of potentially recoverable hydrocarbon deposits during the implementation of the development project. Uncertainty is associated with geological volumes and hydrocarbon recovery rates. The range of uncertainty of recoverable and potentially recoverable volumes of hydrocarbons can be expressed in terms of deterministic scenarios or the statistical probability distribution method (Fig. 2.3).

Hydrocarbon reserves in the "Proved", "Probable" and "Possible" categories are referred to as 1P/2P/3P. Stocks are part of the SPE-PRMS classification and the referencing criteria provided for stocks can, as it is on many occasions, be equally used to Conditional and Prospective resources, provided they satisfy boundaries for discovery or development.

For Conditional Resources, common summing terms are designated, respectively, as 1C/2C/3C. Similar set of definitions of the smallest/optimal/highest score

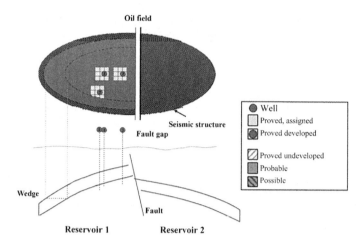

Figure 2.4 Field example of reserves assignation to various categories in line with referencing criteria and for the field development status.

Figure 2.5 Systematic approach for reserve assessment.

apply to Prospective resources. An example how this can be applied to the oil field is presented on Fig. 2.4.

2.2.1 Basic principles of SPE-PRMS classification in the assessment of hydrocarbon reserves and resources

The process of estimating reserves and resources is carried out on the basis of drawing up a project for developing hydrocarbon deposits. The project contains: plans for the hydrocarbon amount assessment as initial geological reserves and resources, identification of the amount of recoverable reserves for each sub-project, project classification based on its current stage and the likelihood of project reaching industrial significance.

As project develops, the categorization of reserves and resources, conducted earlier, is refined. This is done on the basis of careful data analysis and can be systematized as presented at Fig. 2.5.

Below are few definition used to implement the systematic approach.

Reservoir − any natural volume of accumulated oil and/or gas in a trap. Gas, oil and water in the deposits are separated to the great extend under the influence of gravitational forces. Usually gas and oil occupy the upper part of the trap, and water fills the rest of the reservoir. In general, the conditions of the occurrence of oil and gas in the deposits are determined by the hypersometric position of the oil-water (OVC), gas-water (GVK) and gas-oil (GNK) contacts.

Any open or potential reservoir of oil or gas can be the subject of several separate and independent projects at different stages of exploration and/or development. Thus, the same reservoir may have recoverable quantities of hydrocarbons belonging simultaneously to several classes of reserves or resources development projects. The type and quantities of initial geological reserves and hydrocarbon resources are determined from a database based on geological, geophysical and field information. For example, the database will contain: reservoir height, oil, gas, water-oil, gas-oil and gas-bearing zones location, oil-and-gas-saturated thickness and residual oil and gas; reservoir rocks properties and their variations (both laterally and vertically); the initial formation pressure and temperature. Also, the influence of all above constituencies on the final oil and gas yield.

Project − a link between the oil or gas deposit and the decision-making process regarding its development. The implementation of a specific reservoir development project is characterized by a unique hydrocarbon production profile. The profile is unique for only this project and takes into account the time of development and finance dynamics. The project is the main element of this classification, and the net recoverable reserves and resources are additional quantities of hydrocarbons produced during the implementation of each project. Profiles of hydrocarbon production are calculated before the technical, economic or contractual development period. Production profiles define the value of the countable recoverable reserves and resources, and the money flows related to the project. The ratio of NCDs (initial recoverable reserves) to Total initial geological reserves and resources determines the amount of final oil and gas output for a single development project. The implementation of one project may involve the development of several deposits or several projects. All of which can be implemented on one deposit. Not all technically feasible development projects will be industrial. The industrial profitability of a development project depends on the set of conditions forecasted for the project implementation period. These conditions are determined by technological, economic, legal, environmental, social and political factors. Although economic factors can be condensed to projected costs and prices for the outputs, they are also, in the best, influenced by the market conditions, the development of transport and processing infrastructure, taxation conditions and taxes themselves.

Property − legal ownership. Each item, including fields, deposits and infrastructure items, can have a distinctive legal status, which is set by incorporated rights and obligations, including tax conditions. Specific contractual rights and obligations, including financial conditions, allow to define the share of each project participant in the total production (substantiation of ownership), as well as the share of investments, costs and revenues for each development project and deposits. A single property may include several deposits/it is also possible that one deposit may be divided into several properties. The property can be both open and unopened.

2.3 Reserve calculation methods

Initial geological reserves of oil/gas in the reservoir refer to complete hydrocarbon amount in a reservoir and have various names and linked to this abbreviations: Original Oil In Place (OOIP); Oil In Place (OIP); Stock-Tank Original Oil In Place (STOOIP); Original Gas In Place (OGIP). As volumetric amount of oil depends on temperature and pressure one needs to pay attention to the terms meaning.

Methods for calculating reserves (resources) can be grouped into two sets — Deterministic Methods and Probabilistic Methods.

Deterministic Methods — the first set, are based on the existing datasets of geological, geophysical, engineering and economic data. With this approach, for calculating the stocks, single values of each parameter, such as area, porosity, thickness, etc. are used. The result of the reserve calculation in this approach is a single value for the reserve.

Not all parameters in a reservoir are usually known. In this case Probabilistic Methods are used. The methods take into account the uncertainty of geological, geophysical, technical and economic parameters. All parameters are taken with the entire range of possible variations for each initial parameter and might have also time related variables. Stocks are calculated using continuous distribution curves, which are usually obtained using Monte Carlo simulation software. This approach is quite ubiquitous stock calculation.

It is possible, and widely used in practice, to combine two methods as some parameters can be well known and others need to be estimated. Combined method leads to more accurate reserve estimates.

Estimation of reserves and resources at the stages preceding the discovery, or at the stages of the early stage of development, is usually carried out on equivalent, or similar, fields. After the start of production, as data on changes in sampling rates and pressures become available, performance assessment methods may be applied. An increase in amount of information about the reservoir and new data lead to a narrowing of the range of estimates, e.g. level of certainty increases.

2.3.1 Analogy methods

The method of analogies is widely used as a limited amount of initial data for assessing reserves (resources) at the initial stages of exploration and early development of deposits is available. The basis of the method is a statement (an assumption) about the comparability of the properties of reservoir rocks, fluids, thermodynamic (temperature, pressure, phases) and other parameters affecting the assessment of the recoverable reserves in the reservoirs under consideration to the formations-analogues. By selecting an appropriate analogue, the development indicators of which were obtained through the implementation of a similar project (by type and number of wells, size of the production grid and flow intensification methods), you can make a forecast of the dynamics of production of the reservoir in question. The analogue layers are selected by a number of parameters such as depth, reservoir pressure, temperature, displacement mode, initial phase composition, specific fluid density, reservoir size, total and effective oil-saturated thickness, sandstone ratio, lithological composition, formation homogeneity, porosity coefficient, permeability coefficient and technological scheme of development. Similar strata should also be close in parameters to the reservoir in question in relation to geological processes that influenced their formation, including sedimentation, diagenesis, dynamics of changes in reservoir pressure and temperature, chemical composition and mechanical properties of rocks, as well as tectonic deformations.

2.3.2 Volumetric methods

The essence of the volumetric method is to determine the volume or mass (which we obtain by multiplying OOIP (m^3) by the average density of oil under standard conditions) of oil in saturated volumes of reservoir rocks void space.

Volumetric methods are based on the use of geological data. The estimation process starts with accounting for the properties of reservoir rocks and their liquids saturation. The model later develops by determination of the recoverable hydrocarbons, which may be extracted as a result of the development project. Factors affecting the accuracy of estimating initial geological reserves include:

- Geometric shape of the deposit and contours of the trap, affecting the total volume of reservoir rocks.
- Geological characteristics that define the volume of the pore space
- Reservoir properties related to open porosity and residual water saturation

To calculate the initial geological reserves, as a rule, the average coefficient of sandiness, porosity and fluid saturation are used. In the process of calculations, it is also necessary to make assumptions about the natural mode of the reservoir, which largely determines the nature of the displacement of hydrocarbons in the porous medium of the reservoirs. The assessment of recoverable reserves (resources) should reflect the relevant uncertainties not only in relation to the initial geological reserves, but also the recovery factor RF in accordance with the specific development project of the reservoir.

Original Oil In Place is then calculated by:

$$\text{OOIP} \left(m^3\right) = V * \phi * (1 - S_w)/B, \tag{2.3}$$

where V is trap volume, ϕ is coefficient of open porosity, S_w is residual water saturation, B is so named volumetric coefficient which adjust oil volume at the reservoir conditions to the oil volume on the surface after volatile part of the hydrocarbons (essentially gas) and other gases leave the oil.

The rock volume is obtained by multiplying the horizontal projection of the area of oil deposits (A) by the average value of the vertical effective oil saturated formation thickness (h). Only pores (voids) in the rock contain reservoir phases and, the, mobile phases can be only extracted from connected pores. In order to account for this we ought to use open porosity coefficient (Ø) and then also account for water which will be left in pores (voids) after oil extraction.

In the void space of reservoir rocks oil contains dissolved gas. To bring the volume of reservoir oil to the volume of oil degassed under standard conditions, the average value of the conversion factor B is used.

2.3.3 Decline curve analysis

The production decline curve is a plot of production (flow rate) of a well versus time. This dependence is usually plotted on semi-log paper. When data stacks on a straight line, it is extrapolated by an "exponential fall curve" with a constant fall rate.

Analysis of changes in the rate of extraction (production) and the phase composition of the produced hydrocarbons values are jointly used as the information source, which allows to define the value of final recoverable reserves and resources.

The change dynamics in operational performance (such as gas factor, water-oil factor, gas-condensate factor, bottom hole/wellhead pressure) can be extrapolated to the limit of economic profitability of reservoir development.

When data fit on a concave curve, it is extrapolated on a "hyperbolic dip curve". A special case of a hyperbolic dip curve is called the "harmonic dip curve". The most common dependence of the fall curve is a curve with a constant fall rate (e.g. exponential) (Fig. 2.6).

Production analysis curves have many advantages:

1. The data are easy to collect;
2. The curves are easy to plot;
3. There is clear time dependence;
4. The curves are easy to analyze and fit.

If the external conditions which influence the production do not vary by external stimulus then production curve extrapolation is quite reliable for predicting future production volumes.

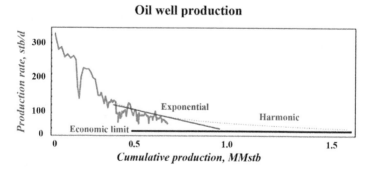

Figure 2.6 Well production volume. Exponential and harmonic extrapolations.

Table 2.2 Main equations for exponential decline production analysis.

Production rate	$q_f = q_i \times e^{-Dn(t)}$
Accumulated (produced) volume	$N_p = (q_i - q_f)/D_n$
Production decline	$D_n = -\ln(1 - D_e)$ и $D_e = (q_i - q_f)/q_i$
Effective decline	$D_e = 1 - e^{-Dn(t)}$
Production time	$t = \ln(q_i - q_f)/D_n$

In the half-logarithmic coordinates can be expressed as:

$qf = qi^*e - D_n \, {}^*t$, where: q — production (flow rate) of the well at time t, bar/day; qi — initial production (flow rate), bar/day; D_n — nominal exponential rate of production decline, 1/day; t — time, day.

In the Table 2.2 the relevant equations are shown.

For deposits at a later stage of operation, production forecast can be defined with an accuracy sufficient to substantially narrow the range of uncertainty on technical risk factors. In such cases, the 2P "best estimate" scenario can also be used to make production forecasts for the 1P and 3P assessment scenarios. Nevertheless, the economic risk factors can remain at a fairly high level, significantly affecting the duration of the project, which should be considered when categorizing reserves and resources.

As oil more and more oil fields mature and there is an increase in the number of previously judged unproductive wells being put into operation, there is currently a tendency towards analysis of the fall rate as proportional to production (hyperbolic and harmonic dependencies). Although some wells have such trends, extrapolation of the dip curve by a hyperbolic or harmonic law should be used with care as excessive use of such dependence in calculation methods can lead to an overestimation of recoverable reserves.

In order to obtain reliable results, when using performance analysis, one, most definitely, requires a stable operating conditions for a sufficient period of time after the start of the production in the established drainage zones. When assessing recoverable reserves and resources, such complicating factors affecting the dynamics of production change as reservoir rock properties, fluid properties, unsteady and steady-state flow ratio, operating conditions, well interaction effects, and depletion mechanisms should be taken into account. In the early stages of production, there may be significant uncertainty in relation to both final operational indicators and industrial factors that affect the duration of a reservoir development project.

2.3.4 Material balance

The material balance method is designed to assess the initial and recoverable reserves (resources) by analyzing the dynamics of pressure changes in the reservoir as fluid is withdrawn from it. The best application of this method is achieved with the availability of the required data points (places where data are taken) of reservoir pressure, the physicochemical properties of fluids at initial reservoir conditions and the full amount of data on the dynamics of hydrocarbon production and water injection into the reservoirs.

When developing highly permeable, homogeneous reservoirs of gas deposits and operating in the natural depletion mode, the reserves assessment carried out by the material balance method can provide high accuracy in determining the volume of final recoverable reserves at various depletion pressure values. In more complex cases, such as water breakthrough into the production wells, complex geological heterogeneity of the reservoir, carbon multiphase deposits, complex strata structures with low permeability reliance on only material balance method can lead to inaccurate estimates. Great care should be taken in assessing the uncertainty factors associated with a specific reserve development project. It is necessary to take into account the features of the geological structure of the reservoir, as well as the dynamics of changes in reservoir pressure during production.

The basis of the material balance methods is to postulate for the oil reserve that there is a fixed amount (volume) of hydrocarbons. This amount, Q_{HO}, is equal to the sum of amounts already extracted during the development, Q_n, and the amount still remaining in the reservoir, Q_{rem}. The total value is constant at any time of development:

$$Q_{HO} = Q_n + Q_{rem} = const \qquad (2.4)$$

Differences in the nature of the indicators, both geological conditions and conditions for the development of each specific deposit, require an specifically developed approach to the compilation of the material balance equation in definition of initial reserves and produced hydrocarbons.

The most reliable results in calculating reserves using the material balance methods can be obtained when the values of the measured and estimated parameters (the reservoir pressure is very important) are accounted in the values in the equation in the formula for calculating reserves. The parameters should be representative and characterise the entire reservoir as a whole.

2.3.5 Reservoir modeling (simulation)

An improved form of the material balance method can be developed to be more accurate by computer modeling of the reservoir. This modeling will provide highly accurate estimation of the reservoir response during the application of a specific technological development schemes. The model's accuracy is directly dependant on the reliability of the initial data, such as reservoir rock properties, geometric reservoir shape, relative phase permeability functions, fluids properties, etc.

Geological and hydrodynamic models represent the reservoir as a set of three-dimensional interconnected cells or "reservoirs" and use the initial data based on the bulk method or the method of production dynamics. Models then can be used to calculate the initial geological volumes by taking into account the adaptation of the dynamics of changes in reservoir pressure, hydrocarbon production by wells or deposits, forecasting the expected flow rates for various development project strategies.

Methods of geological and hydrodynamic simulation (e.g. modeling) integrate all the stages of geological, geophysical, petrophysical and industrial exploration of the formation and its reserves. The already developed modeling software has been widely used and applied to the reservoirs where a more accurate idea of the spatial distribution of fluid and reservoir properties is required, both in terms of area and deposit.

The development of new software packages continues. The modeling and software development encircle the entire spectrum of research and production for a field resource potential. The modeling is ranging from calculating reserve volumes to assess development risks to optimizing various reserve development scenarios.

2.4 Oil recovery factor

The development of an oil and gas field is a continuous process involving various stages of exploration and development. Each stage is characterized by a certain degree of geological, geophysical and field explorations of the formation. All this, in turn, defines the methodological approach to the calculation of reserves and, in turn, affects the reliability of reserve calculation. It is well known that the reserve assessment accuracy decreases in absence of actual geological and geophysical information at the initial stage of exploration and especially in situations of significant variability and heterogeneity of geological and field parameters of hydrocarbon deposits.

It should be noted, that in defining the strategy for the development of hydrocarbon deposits one needs to take into the account the initial geological hydrocarbons reserves. Environmental protection should be taken into the account together with the amounts of hydrocarbons (recoverable reserves) which can be produced on the basis of geological, technological and technical capabilities of the development project.

The recoverable part of reserves (oil, gas, and condensate) is defined by introducing the recovery factor (RF). For oil it is called oil recovery coefficient, for gas and condensate, respectively, gas recovery coefficient and condensate recovery.

The oil recovery factor (at the particular time) is defined as the ratio of the value of recovered reserves (Q_{rec}), that is, already extracted to the surface, to the value of geological reserves (Q_{geol}) located in the depths of the deposit:

$$RF = Q_{rec}/Q_{geol} \tag{2.5}$$

Oil recovery from reservoirs (ORF) has a significant effect on the projected (planned) volumes of capital investments in the process of developing hydrocarbon deposits, as well as on the likely estimate of the growth of industrially recoverable reserves. The accuracy of oil recovery determination has a direct impact on the assessment of residual reserves, the effectiveness of the development systems used, the prospects and scale of the introduction of new hydrocarbon production technologies, and the timing of the commercial development of deposits.

An increase in the oil recovery coefficient of at least 0.01 (a very minor amount) in its economic significance is equivalent to the discovery and commissioning of a new field in industrial development. Solving the problem of achieving the highest possible oil recovery ratio is one of the most urgent objectives of petroleum geology and the development of oil and gas fields.

The oil recovery factor calculated over the entire development period is referred to as the ultimate coefficient (Ultimate Recovery Factor — URF). Calculated over a specific time period from the start of development to the evaluation date — the current RF.

Currently, several methods for calculating the final (expected) RF are widely used:

- The method of analogies makes estimations on the basis of the natural reserve oil driving forces (Reservoir Drive Mechanisms);
- Statistical method is built on the basis of obtaining dependencies between the final oil recovery coefficient (RF) and various geological, physical and technological factors;
- Coefficient method finds its roots in determining the values of a number of coefficients when taking into account the geological and field characteristics of some specific oil deposits and features of the planned development project;
- Hydrodynamic method takes technological calculations of few reservoir development scenarios by modeling a specific hydrocarbon deposit.

Based on more than a century of experience in the development of hydrocarbon deposits, it has been established that the oil recovery factors (RF) achieved as a result of the implementation of projects vary in a wide range from 0.05 to 0.75 and higher. The achievable value is mainly related to the application of secondary (methods for maintaining reservoir pressure, i.e. injection of water into the reservoir) and tertiary (new methods of enhanced oil recovery of reservoirs — EOR — enhanced oil recovery) production stimulation methods.

If RF reaches 40—50% or even higher then oil formations are classified as active (for low-viscosity oil in highly permeable reservoirs). If RF not exceeds 10—30% despite using traditional methods to displace oil from the porous medium, then the reserves are called hard-to-recover (usually high viscosity of oil, or low-permeable reservoirs, non-traditional collectors).

The coefficients of oil-gas-condensate recovery are determined on the basis of technological and technical-economic calculations when compiling a project for the development of a specific reservoir.

The magnitude of the projected RF depends on the complex geological, physical and reservoir parameters. When all of this is determined by the lithological composition of reservoirs, the heterogeneity of the structure of the reservoir, permeability and effective oil-saturated power, the manifestation of the deposit natural drives. The methods (technologies) used in the extraction of hydrocarbons also have a direct impact on the achievement of maximum oil recovery ratios. One can account for: producing wells grid density; well location on the reservoir structure; current technical well state; methods and methods of intensifying oil production; implementation of reservoir pressure maintenance systems; the use of methods of influence on the reservoir. The list can go on and on.

The **Method of Analogies** can produce rough estimates which are dependant on the natural operation modes of the reservoirs (e.g. Reserve Drive Mechanisms).

According to the basic physics laws reservoir liquids will move under existing (natural) or intentionally generated forces. Such forces include pressure differential, gravity, capillary and hydrodynamic influences. The lowest pressure in the oil reservoir should be near and inside of production well. We assume that the well has perforations and oil is actively pumped out. If we have the reservoir pressure higher than in the production well and capillary forces are lower than the pressure differential, then the reservoir liquids (we hope that this is mostly oil) will move towards production well and will be extracted.

Under the oil deposits natural drive regime there are historically existing in the reservoir forces which move oil (hydrocarbons) in the pore space of reservoir rocks to the production well perforation holes. The natural reservoir regime has one of the most important effects on the choice of the most economical reservoir development strategy and the best efficiency of reservoir energy utilization.

The driving pressure in the reservoir can be produced by the gravitational force (Gravity Drive), gas cap (Gas Cap Drive), oil elastic expansion or by Solution Gas Drive. Water from existing aquafers can "push" oil from the formation side(s) or from the formation sole (Waterdrive).

On the basis of historical data, modeling and laboratory studies it is accepted that the natural drive recovery coefficient can have the following values:

Gravity drive	≥ 0.5
Gas Cap dive	$0.2-0.4$
Solution Gas Drive	$0.05-0.3$
Waterdrive	$0.35-0.75$

Gravity exhibits itself in various roles. If we pump the oil from the bottom of the formation then oil will keep coming to the production well under gravitational force until we reach the balance between gravity and the capillary forces. We have to assume that in place of removed oil the reservoir is filled with some sort of gas.

Gravity force is also important when we have liquids of different density. Let us assume at this point that we have homogenous oil which has specific gravity lower that water or, in other terms, API is above 10. In this case water will collect at the bottom of the reservoir and, having water supply, will continuously push oil upwards until all reservoir is almost fully filled with water.

On Fig. 2.7 there are examples of combined gravity and water drives. In both cases there are also some gas caps but it is assumed that the pressure in there is not driving force.

Water assisted oil displacement also can be observed as related to water properties itself phenomena. In this case water expands during the initial pressure drop, so

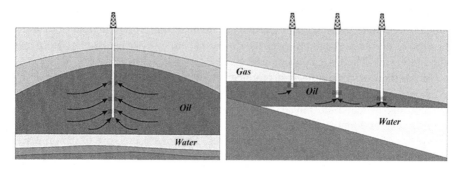

Figure 2.7 Combination of gravity and waterdrive in the horizontal and inclined liquid containing strata.

named elastic-pressure mode. Elastic mode — the mode of reservoir operation in the process of reducing the pressure in the reservoir. The reservoir energy exhibits itself in the form of an elastic expansion of the reservoir fluid and rock. Compared with the water-pressure mode, the elastic-pressure mode of the reservoir operation is less effective. The oil recovery coefficient in this mode of reservoir development can be between 0.3 and 0.6.

Water drive mode (Waterdrive) — is an active water drive mode. It is the most effective mode in which oil moves in the reservoir to the production wells under the pressure of contour (or bottom) water. In this mode it is possible to extract 35–75% of OOIP. Sometimes the extraction is even higher. In average the recovery factor for Waterdrive is thought to be at 50%.

It is also possible that the gas above the oil in the formation, so named gas cap, has some significant pressure. As we pump the oil from the reservoir the oil level lowers and gas expands but still presses oil downwards towards out production well perforation. This situation is named gas cap drive. The recovery factor in this case is within 0.2–0.3 and is taken in average as 0.3. For most economical exploitation it is recommended to maintain gas cap pressure by injecting natural gas. In some cases it was possible to do natural gas injection even after significant depletion of reservoir pressure and restore or initiate gas drive.

As we start production the pressure in the reservoir might drop. This leads to slight oil expansion as an elastic body and this expansion for a while will keep the pressure and oil flowing to the production well. It is also possible that during pressure drop the liquid passes the bubble point pressure, e.g. some dissolved gas will start to form bubbles. This gas separation can continue for a while and maintain some almost stable, below the original, pressure in the formation. This is so named solution gas drive.

The drawings on Fig. 2.8 reflect on the above diving mechanisms.

In almost all cases there is so named mixed natural drive. It means that two or more of the above mechanism are present and are active or will activate as the pressure in the reservoir change.

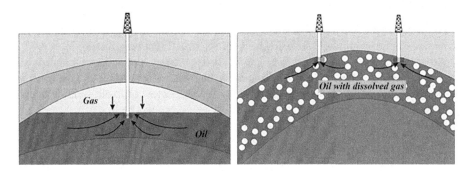

Figure 2.8 Gas cap and solution gas drive scenarios.

The natural reservoir regime has one of the most important influences on the choice of rational, most economical, reservoir development systems and the efficiency of utilization of reservoir energy in order to achieve the maximum ratio of oil (carbohydrate) recovery.

At present, the justification of RF is conducted, as a rule, on geological and hydrodynamic models, taking into the account geological, technical, technological and even economic risks. The geological part of the model includes a three-dimensional geological representation of the deposit in the form of a digital model. The hydrodynamic part includes complete information on the operation of production and injection wells (perforation intervals, commissioning, work history in time, etc.).

Reliability of geological and hydrodynamic modeling is also achieved by their adaptation to the past actual period of development of the reservoir (History match). By simulating filtration processes in the three-dimensional space of reservoir rocks and displacing oil (working agent) to the bottom of production wells, a special model of the development project is created using computer programs to predict the change in the state of the reservoir behavior at any time. The result is a calculation of the design indicators of the development, both by the years of operation, and for individual periods (10, 20, 40 years) of time, or until the end of development.

The value of the final oil recovery coefficient RF is determined for several variants of the design of the development system, which represents the overall optimization task of choosing the best design variant that ensures the maximum oil recovery ratio.

Two different approaches can be taken as the basis for optimization: the achievement of the technically possible maximum oil recovery coefficient, i.e. getting the highest RF, or getting the most profit from the development project implementation process. The priority of assessing the oil recovery ratio, taking into account the economics of a development project, requires that the current oil market, tax legislation, etc. be taken into account in the calculations of the current market situation and, ultimately, profit margin.

As in any enterprise, the economic factor has become dominant and the priority of obtaining optimal profit. Thus, modeling in most cases required to focus on the economic component of the oil recovery coefficient RF recovery factor.

All oil and gas companies have an interest in development of enhanced oil recovery methods. Well focused research is being conducted which is aimed at finding a scientifically based approach to choosing the most efficient field development technologies.

2.4.1   Regulatory bodies and frameworks

- PRMS – Petroleum Resources Management System. Hydrocarbon Reserves and Resources Management System
- SPE – Society of Petroleum Engineers. Classification of the Society of Petroleum Engineers
- WPC – World Petroleum Council
- World Petroleum Congress
- AAPG – American Association of Petroleum Geologists
- SPEE – Society of Petroleum Evaluation Engineers.
- SEC – Securities and Exchange Commission
- UNFC – United Nations Framework Classification
- SORN – United Kingdom Statement of Recommended Practice
- CSA – Canadian Security Administrators
- Russian Ministry of Natural Resources
- PRO – China Petroleum Reserves Office
- NPD – Norwegian Petroleum Directorate
- USGS – United States Geological Survey
- For more information please refer to SPE Oil and Gas Reserves Committee "Mapping" Subcommittee Final Report (October 2005) at http://www.spe.org/spe/jsp/basic/0,1104_4544118,00.html

Basic physical and chemical concepts

3

Abstract

In order to extract oil from a formation we need to be able to displace reservoir fluids into the production well and move them to the surface. On the first place we need to have or be able to create driving force, either utilize gravity or to have pressure differential, which might eventually move our fluids, hopefully containing hydrocarbons, into the production well. On the second place, for the fluid to flow it needs the interconnected channels (openings, interconnected pores) up to the production well. How fast the fluid flow depends on its viscosity, which is an indication of the liquid "internal friction", size and concentration of interconnected pores and physical interactions between fluids and the formation stone.

In order to extract oil from a formation we need to be able to displace reservoir fluids into the production well and move them to the surface. On the first place we need to have or be able to create driving force, either utilize gravity or to have pressure differential, which might eventually move our fluids, hopefully containing hydrocarbons, into the production well. On the second place, for the fluid to flow it needs the interconnected channels (openings, interconnected pores) up to the production well. How fast the fluid flow depends on its viscosity, which is an indication of the liquid "internal friction", size and concentration of interconnected pores and physical interactions between fluids and the formation stone. One can immediately see that the reservoir fluids flow is affected by many variables and is very complex process. Flow analysis is not made easier by the fact that many fluids properties depend on temperature, pressure and even speed of flow. On the top of all mentioned, there are significant heterogeneities in the formation, which include both — spatially irregular stone and reservoir fluids properties. The useful experiment to analyze the flow should account for the significant part of the above and predict how the flow would change with the deviation of foreseeable inhomogeneity which can be encountered during extraction process.

The painted extraction picture is so enormously complex that the full analysis is impossible. In the past the complexity was taken into the account very lightly or ignored at all. Nowadays, fortunately enough, having much better understanding of basic processes and computing power, we can make some assumptions, state some parameters as constants and neglect some processes as insignificant. In the bold terms all this will depend how much it will cost us to reasonably understand processes or influences of various eventualities. And now much more we can get (hydrocarbons and profit) if we would care about everything and try our best. It

Primer on Enhanced Oil Recovery. DOI: https://doi.org/10.1016/B978-0-12-817632-0.00003-7

goes without saying that if we know and control more parameters we will get more oil but the question is what price we will pay for the knowledge, exerted influence and what the profit will be.

For the future clarity we will provide few definitions here. One needs to take into the account that in many cases, and this book is no exception, the definitions are linked to an application area. Here we will provide the definitions which are more suitable for oil and gas industry and also sometimes use more ubiquitous terms.

The matter exists in different, so named, states. Each state has set of physical parameters. Regarding hydrocarbon extraction, the matter (hydrocarbons) can be in solid, liquid or gaseous state. In some cases we can have vapor which can be a mixture of gaseous and liquid state (small droplets of liquid in gas) in the state of thermodynamics equilibrium.

In general description fluid is a matter which does not have permanent shape. It is also said that a fluid deforms under shear stress. A fluid flow is a special case of deformation. In principle fluid term includes liquids, gases and supercritical fluids. For the purpose of this book we will only concentrate and talk about liquids.

In order for fluid, liquid from this point, to flow one needs to apply and maintain pressure (differential, e.g. sheer force). If there is no sheer force liquid flow will cease (albeit only after inertia energy of the motion has been zeroed). This is due to the fact that liquid "resists" deformation. One can observe that various layers of liquid during deformation move with the different speed. Interaction of those layers between themselves appears like internal friction. The value of this "internal friction" is represented by a value of viscosity. Liquid almost always, except few very special case, interacts with the surfaces which defines liquid shape (a pipe wall, for instance). For the majority of liquids, when the liquid moves the liquid layers very near solid surface has much smaller speed than the liquid layers in the middle of the flow (we assume here that the liquid moves in the pipe and that the liquid moves relatively slowly so that the flow is laminar). In accordance with the above liquid layer will start to experience "internal friction" and we will observe the viscosity. More accurately, in this case, it is so named dynamic viscosity.

The dynamic viscosity can be defined as this

$$\mu = \frac{\sigma}{\gamma} = \frac{F}{A} / \partial v / \partial y$$

where μ is viscosity. It is obtained then as a ratio of stress to gradient of flow speed. Given that the speed of movement near the pipe wall is very small, then small gradient of speed means that all liquid moves slowly in all volume and the viscosity is high. We can say that liquid with high viscosity flows slowly! Actual laws of liquid flow through a pipe are quite complex. Many parameters need to be taken into the account but the fact remains − viscous fluids flow slowly and one needs to maintain high pressure (differential) to maintain the flow.

Table 3.1 Dynamic viscosity at 20 °C in cP (equal to mPI).

Water (pure)	1
Seawater	1.1
Ethanol	1.1
Kerosene	1.6
Various Brents	5−7
West Texas Intermediate	Around 5
Boscan (10 °API)	>50,000

In System International (SI) the unit if viscosity is Poiseuile (PI). In Centimeter-Gramm-Second system (CGS system) the unit of viscosity is Poise (P). Unfortunately a single unit is a bit too big for everyday use, so we instead use one hundred of it − centiPoise (cP). It is worth remembering that 1 P = 0.1 PI. All the below assumes that the viscosity of liquid is not dependant on the sped of flow (which is not true in all cases) − we name this class of liquids − Newtonian fluids. The lower the viscosity the more fluid will move through a pipe at the same pressure. It is much easier and cheaper to pump a liquid with low viscosity (Table 3.1).

Viscosity of an oil is linked to the chemical composition, API, temperature, pressure and amount of dissolved gas in oil. Dependence on chemical composition is quite complex and cannot be presented in simple terms. As the gravity of oil (°API value) goes down the viscosity increases very steeply. Dissolved gas reduces oil viscosity. It means that if the pressure in the reservoir drops and passes the bubble point and gas separate from the oil and the oil will become more viscous. Below bubble point oil viscosity increases with the pressure reduction, above bubble point pressure the oil viscosity increases with the pressure.

The temperature has dramatic effect on viscosity. Even simple liquids like water will have big changes of viscosity as temperature rises. For example water dynamic viscosity will be reduced by the factor of almost four as we increase temperature from 20 to 100 °C. It is not unknown for the crude to reduce viscosity two times for every 5−10 °C during increase in temperature. It means that increase of oil temperature by 50 degrees in many cases can reduce oil viscosity more than 100 times. High temperature in a reservoir is good for low viscosity and easy liquids movement.

A phase is an amount of matter with uniform chemical content and consistent molecules (atoms) structural arrangements. A phase has the set of well-defined properties. Knowing the phase one can know its properties. The opposite is also true.

The phase diagram is a drawing which links a matter properties with some thermodynamic parameters. It is probably easier to understand complex diagrams starting with a simple one, known and chemically simple liquid, water. As an example on Fig. 3.1 there is a simplified and slightly imprecise water phase diagram. There are four water states − solid, liquid, vapor and supercritical fluid. Lines between the states represent the equilibrium values. In the triple point liquid, vapor and solid (ice) exist in thermodynamic equilibrium. By definition at the sea level atmospheric pressure this is happening at approximately 0 °C or at around 273 K. Above the

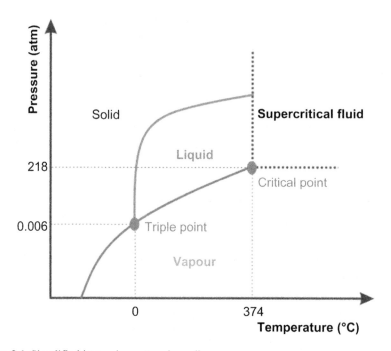

Figure 3.1 Simplified imprecise water phase diagram.

critical point [pressure above approximately 218 atm (22 MPa)] and temperature above 374 °C (647 K) vapor and liquid are undistinguishable. This is so named supercritical fluid and properties of it are outside of this book interest

Vapor is the material in a gas phase. Vapor exists below critical temperature and can condense in liquid or solid state depending on the pressure. Vapor can be formed from liquid (by evaporation or boiling) and from ice (by sublimation). Transformation from liquid to vapor requires significant energy. This energy is released when the vapor condenses back into water. This energy is approximately six time bigger than the energy to heat water from 20 to 100 °C. This is exactly why water vapor is so efficient to deliver thermal energy to an oil formation.

In many cases water vapor term is interchangeably used with water steam term. This is only correct if the steam is dry. In this case water gas (vapor, steam) is fully transparent and it has lower density than air. If the steam is not dry then it contains water droplets and it is translucent. Amount of thermal energy in a stem is significantly affected by the temperature and amount of water in droplets. Ratio of water droplets in the steam defines steam quality.

The bubble point is a temperature and pressure at which vapor starts to separate from the liquid in a form of bubbles. At constant and high enough temperature this will happen if the pressure is reduced. In everyday life this is can be observed when a can with fizzy drink is opened (the pressure is reduced) and the liquid suddenly becomes full of bubbles. The opposite is also true, if the pressure has increased the vapor will start actively dissolving in the liquid. The pressure when all vapor will dissolve is a dew point.

When a chemical compound is in a liquid or solid state the attraction of molecules to each other can be higher than the other molecular forces (repulsion forces plus thermal motion). When two chemically different matters brought in contact the molecules at the boarder layer start to interact. The fluid tends to minimize the surface area due to the molecules mutual attraction and will form a droplet. The liquid resists any surface area alteration, in the scientific terms some energy needs to be supplied to increase surface area. As the result we can say that there is a force which is named surface tension. If we want to increase the surface area we will need to work against this surface tension. For instance, for water in air at room temperature (say 20 °C) the surface tension is at around 74 mN/m. If we talk about the area than we will need to talk about the surface energy.

Mixing or blending of different compounds is extremely important for the hydrocarbon extraction. Without going into much of a detail we will state that gasses do mix without too much difficulties. Gasses and liquids, liquids and liquids, liquids and solids will mix only in certain combinations and only in the defined conditions.

If two phases (let's say − two liquids) can be mixed at all ratios and create only one phase then these two phases (liquids) are referred to as miscible. In this case there is no interface in the volume which will separate two phases. It looks easy but it is not. For instance, water and alcohols at temperatures below 0 °C will separate − water will freeze and alcohol will remain a liquid until it freezes too at even lower temperature.

If at certain temperature and pressure two liquids (or gas and liquid) mix immediately as they brought into a contact (e.g. they produce a single phase) then this named First-Contact Miscibility (FCM). In this case two phases already mix at the lowest possible pressure. Miscibility increases with pressure and at high enough pressure almost everything makes a single phase. At the same time high pressure is expensive to produce, implement and sustain. From this point of view it is important for the given reservoir conditions to know the minimum miscibility pressure. At the multiple contact miscibility (MCM) two phases first exchange components. By exchange process the phases obtain undistinguishable properties and create a single phase. It is widely accepted that the pressure for MCM is lower than for FCM. Volatile components in oil play an important role in defining the oil properties. As the result, miscibility of live oil (oil with volatile components) and dead oil (without volatile components) do mix differently with other fluids. The general rule is that the solubility increases with the pressure and decreases with temperature. Interplay between influences of oil properties, pressure and temperature on miscibility is very complex and needs laboratory data for proper assessment from all points of view.

If two liquids do not mix (do not create a continuous single phase system) it is still possible to create a solution containing both phases. It is generally known as emulsion (colloidal system is more encompassing term). Emulsion preparation, properties and behavior are very complex subjects and we will just touch on them here and provide very simplified description as an introduction.

An emulsion is a dispersive system when one phase is spread in another phase in form of droplets. Simple emulsions are not stable and over time will separate into

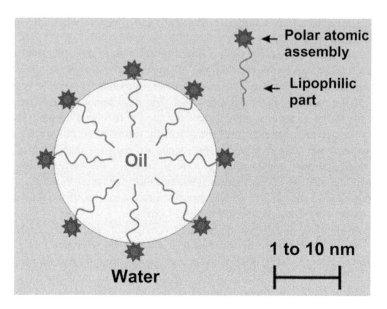

Figure 3.2 Oil micelle in water.

two continuous phases with a single interface. One needs to bear in mind that it is possible to create various emulsion even for the same pair of liquids. For instance, both oil-in-water and water-in-oil are possible. Even more complex emulsions are also possible. The properties of an emulsion are very much depend on size and concentration of droplets. As a simple emulsion is not stable, many things can shorten it lifetime. Temperature change, pH change, mechanical interaction (pumping for instance) can reduce an emulsion lifetime and change the emulsion physical properties, in some cases almost instantaneously.

In order to stabilize an emulsion and make it more well dispersed (very small droplets) it is possible to add chemicals known as emulsifiers, also known as surfactants (or detergents). Emulsifiers contain long molecules with one part (polar atomic assembly, sometimes referred as head) easily dissolvable in water (hydrophilic part) and another one (nonpolar atomic assembly, sometimes referred as tail) is easily dissolvable in oil (lipophilic part). The molecules then concentrate on the interface between water and oil and make bridges between two liquids. Fig. 3.2 shows "simple" micelle of oil in water. It is important fact that existing of micelle in the liquid is guided by complex thermodynamics and the form of micelle is guided by the concentration of surfactant. At low surfactant concentrations there could be no micelle assembly. The micelle concentration and shape also depend on surfactant concentration and other external factors. Micelle can have more complicated internal structure with small micelle in a middle of a big micelle.

When the micelle formed they significantly affect liquid viscosity and rheology. On the most basic level of description one can count them as very big molecules. The exact behavior is very complex and is beyond this text remit.

Fluid-rock interaction

4

Abstract

Three material types form an oil reservoir — formation rock, liquids and gases. We hope that the liquids will be just oil and the gases would be just some light hydrocarbons. As it is an oil formation then we should predominantly have rock and hydrocarbons. But as life goes, we will have some water too. All elements interact with each other and before we start oil and gas extraction all of them are in the thermodynamic equilibrium. As we start extraction the equilibrioum is no more and the processes become even more intricate.

Three material types form an oil reservoir — formation rock, liquids and gases. We hope that the liquids will be just oil and the gases would be just some light hydrocarbons. As it is an oil formation then we should predominantly have rock and hydrocarbons. But as life goes, we will have some water too. All elements interact with each other and before we start oil and gas extraction all of them are in the thermodynamic equilibrium. Equilibrium means that all chemical reaction have already finished and any changes have stopped long time ago.

All elements in the reservoir are complex and have many subcomponents. Description of all balances, extraction dynamics and interactions are impossible. Still, it does not mean that we should not try to understand the most important moments and predict with a commercial susses the whole system dynamic during oil extraction. Water usually appears as a relatively simple liquid, but nothing is simple about the connate water. It can be reasonably understood from three main points — water is polar liquid, it contains dissolved salts and water strongly interacts with the rocks. Water dissolves many chemical salts and dynamically interact with the rock. Dynamic interaction means that some salts in form of ions interact with rock and stick to the rock surface. Some minerals from the rock surface dissolve in the water if thermodynamic conditions change. Those processes create absolutely new rock surface which has very different chemical representation and will interact completely differently with water and oil.

Water, oil and gases interact with rock. Let us forget about gases and say that we only have oil, water (two immiscible fluids) and a rock surface. First phenomena we encounter is wettability. Wettability in essence shows how strongly surface interacts with a liquid. The first useful model is represented on Fig. 4.1.

The figure shows interaction (adhesion) between rock and liquids — water and oil. The model contains a droplet of water in oil on a rock surface. In the case a) water/rock interaction is weaker than oil/rock interaction. In this case we will talk

Primer on Enhanced Oil Recovery. DOI: https://doi.org/10.1016/B978-0-12-817632-0.00004-9

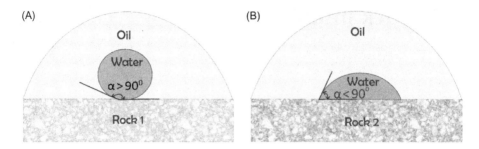

Figure 4.1 Wettability on a rock surface. (A) Oil-wet surface and (B) water-wet surface.

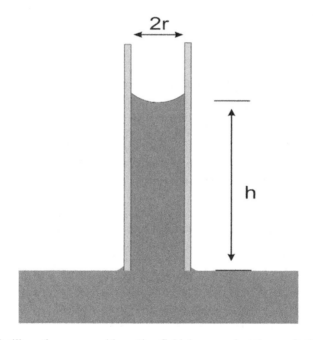

Figure 4.2 Capillary phenomena with wetting fluid drawn against the gravitational force.

about oil-wet rock. In the case b) water/rock interaction is stronger than oil/rock interaction and we talk about water-wet rock. It is possible to measure an angle at the point where water surface meets the rock surface. This is so named contact angle. If the angle is higher than 90° then the rock is oil-wet. If the angle is below 90° then the rock is water-wet. When angle is around 90° we talk about neutral wettability.

The next step on analysis of liquid/rock interaction is capillary phenomena (see Fig. 4.2).

The capillary pressure (this is in the presented picture a pressure which drags the liquid up against the gravitational force) can be expressed as follows:

$$p = \frac{2\gamma cos\alpha}{r}$$

where γ is interfacial tension, α is contact angle and r is the capillary radius. As radius (size) of the capillary gets smaller the pressure is getting bigger. In other words, the wetting phase gets into a small opening in the rock and it is very difficult to get it from there.

In general terms, wetting phase coats the rock surface.

If we talk of rock types wettability then fresh carbonate, quartz and dolomite are water-wet. But life is newer as simple as that — after contact with oil the mentioned minerals can become oil-wet. At the same time, as temperature rises quartz becomes more oil-wet while calcite switches it affiliation to water.

It has been seen that small pores can be water-wet and are filled with water, while the larger pores are oil-wet and filled with oil. In many cases wettability is defined on an ionic level — it all depends on ions on a rock surface. In a slightly bigger scale the wettability depends on surface roughness and grain properties (size and shape). The wetting phase is difficult to displace and this is why wettability is important for oil recovery. In water-wet rock an oil can be trapped in large pores.

The wettability can also change as the chemistry of water (dissolved minerals) is modified. One can see that the wettability is a very variable phenomenon.

What makes life even more difficult is that it is generally agreed that there is no method to measure the wettability reliably.

Reservoir mineralogy and rock-fluid interactions

5

Abstract

The hydrocarbons and connate water are contained in porous rock. Rock fluid interactions, fluid properties and the rock porosity need to be well analysed to enhance oil recovery. As a single phase oil flow through the porous media can be predicted by Darcy's Equation. Unfortunately the simple equation very fast fails in the real multiphase flow. The liquids flow is also affected by the partial saturations and partial permeabilities. Complete liquid extraction from the porous media at the oil field is impossible due to the capillary forces.

An oil field is a combination of hydrocarbon deposits confined to one or several traps. The traps are connected by a common geological structure. The hydrocarbon deposit is a rock with hydrocarbons in the rock pores. The reservoir rock consists of cemented mineral grains. The porous rock with oil can be simply represented as shown in Fig. 5.1.

The grains can have various sizes and shapes. Not all grains would have the same size, mineralogy and origin. Overall grain structure and composition can change in vertical and horizontal directions or in any part of the deposit. This fact is referred as heterogeneity. Consolidated grains inevitably have some spaces between them and the space in an oil formation is filled with oil and connate water. Presence of grains also has another important manifestation — the material with grains and porosity has very high surface area. It is known that for the sandstone surface area can reach 5000 m^2/kg, while for shales it even can be up to 20 times bigger than this and can reach 100,000 m^2/kg. This is an enormous area, a shale cube with a site just around 3.5 m will have bigger surface area than London.

What does not change in principle that in order to contain liquids the rock should have porosity. The porosity is some volume in-between grains, rock voids and cracks. In general, the porosity is defined as an amount of empty volume in the material (rock) expressed in the percent form:

$$\phi = \frac{V_o}{V_r + V_o} \times 100\% \tag{5.1}$$

where V_0 is volume of all pores, V_r is volume of the solid matter. In the sedimentation process porosity depends on the grain shape and size distributions. If the grains have the same size (we assume them spherical and use so named Kepler conjecture)

Primer on Enhanced Oil Recovery. DOI: https://doi.org/10.1016/B978-0-12-817632-0.00005-0

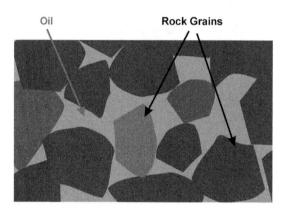

Figure 5.1 Porous rock with oil.

then porosity will be at the maximum level at around 26%. Other grain shapes are widely presented and the maximum porosity can be higher. In general, as grains have in many cases very irregular shape and various formation origins, the porosity can vary in very broad spectrum from 10% to 40% for oil containing geological formations.

In order for oil to move (to enable an oil extraction), the pores should be connected. Otherwise the oil would not be able to flow through the rock towards the production well. On the first place we are interested in the volume of connected pores — so named open porosity. It might happened that not all pores contain oil. In this case one can talk about effective porosity — the volume of connected oil containing pores.

Saturation is defined as a percentage or a fraction of a liquid in the whole pore volume. Essentially it shows how much of a pore volume is occupied by a liquid. Related to water it then shown as k_r and to oil as k_o. The same as porosity, the saturation can be total (related to the total pores volume) and can be effective. The latter is used much more often.

The ability for reservoir liquids to flow is most important for oil recovery. The liquids flow through the chains of connected pores. The flow in two dimension can be easily visualized as a river flow in a map though the series of valleys. In general, this analogy shows that as the whole the liquid flows between two distant points but hardly on the straight line. The flow twists and turns, sometimes makes almost full loops and in some areas flow towards the flow source. The same, but in three dimensions, happens with the flow of liquid through the porous media.

The basic equation which guards a liquid flow (as a single phase) though the porous media has been constructed by French engineer Darcy in 1856. The original equation was produced for water permeation through sand. After all Darcy was a water engineer. The equation is a significant simplification and has many limitations. Nevertheless, it has some applications and helps to understand the basic processes. Adapted to the conventional form the Darcy's Law (Equation) can be expressed in the form:

$$Q = -\frac{kA}{\mu}\frac{dP}{dL}$$

where Q is a flow rate, which is an oil volume in a time unit. Traditionally Q was measured in mL/s. A is area of the opening (this is the whole surface area through which the flow is measured), μ is viscosity, k is permeability. Depending on the measuring system units the permeability can be measured either in square microns (in SI) or milliDarcy (mD) (in CGS). One Darsy is approximately equal to 1 square micron (μm^2). The last term, dP/dL, defines differential of the pressure which invokes the flow.

Usually the conventional oil reservoir rocks are semi-pervious and permeability is roughly in the range between 100 and 10000 milliDarcy (from 0.1 to 10 μm^2). The high values are favorable for oil production as for the same pressure differential the quantity of flowing oil is higher. Highly fractured rock can have permeability up to 10^5 Darcy. On the low of permeability are dense clays and granite with permeability in the region 10^{-7} Darcy.

The Darcy's law deals with a single liquid flow through a porous medium. Reservoir fluids are mixtures of hydrocarbons (oil and gas) and connate water. Ability of rock to allow a single phase flow is referred as an absolute permeability. In the case of two or more phases there are effective, so named partial, permeabilities for each phase. In this case each effective permeability is lower than the absolute permeability and the sum of all effective permeabilities cannot be bigger than the absolute permeability. All permeabilities are affected by the rock saturations with all presented phases, rock wettability and speed of liquid flows. This makes liquid flow in the reservoir rock very complex and the only reliable and accurate way to assess the oil flow is to do measurements on the rock cores in a laboratory under reservoir conditions.

It is clear at this point that reservoir fluids flow through openings of different sizes and the flow is affected by viscosity and interfacial tension. In order to reflect on this a dimensionless value, named a capillary number N_c, is defined in oil industry as

$$N_c = \mu/\sigma$$

In small pores the capillary number is very small, usually below 10^{-6}. While in the cracks and pipe-like flow it is approaching 1. In the first case the flow is defined by the interfacial forces, while in the second case flow is mostly defined by the viscosity. It has been shown that an increment in capillary number in the in the formation has benefits in reduction of the residual oil saturation. The approach is then quite obvious − reduction of an interfacial tension leads to the better oil recovery.

More phenomena makes liquids flow through the porous media a very intricate process. Firstly, it is the presence of so named fines. The fines are particles of smaller size than the main mineral grains and are not cemented into the structure. The fines can have different mineralogical origin than the dominant formation rock or can be produced as the main rock gets fractured or gets less consolidated. The fines also can be produced from some rock types due to chemical changes during oil extraction.

An example of fines is shown on Fig. 5.2. In this case the fines have other the main rock origin (they can be a clay, for instance). The fines easily move with the liquid flow and eventually block some pores interconnections (throats). The faster is the flow, the faster the blockage can be established. Presence of especially clay fines due to their plate-shape, ability to swell and complex influences on liquid rheology represent one of the highest challenge during oil extraction.

The pores in an oil reservoir contain reservoir fluids − hydrocarbons (oil) and some connate water. At this point we need to introduce a term of saturation. Saturation (S) is defined as a fraction of pores volume occupied by the reservoir liquids. Usually saturation divided into saturation of water, nominated as S_w, and saturation of oil S_o. During oil extraction both of those saturations change. The effect of saturation on reservoir behavior will be discussed a bit later.

The minerals in formation can be water-wet or oil-wet. In means that in many cases the wetting phase will coat an entire grain with thin layer as shown in Fig. 5.3.

Type of wetting affect dynamics of liquid flow during oil extraction and the reservoir final saturations. In many cases not only rock type defines the wettability in the dynamic situation but also the past saturation. Establishes layers of water on the grain surface within surrounding of oil only can be well developed if the pores in the past were water saturated. Small pores tend to be predominantly occupied by the wetting phase.

Saturation and capillary pressure behaves very special ways when the liquids drained from the pores. A schematic example is presented on the Fig. 5.4. In this

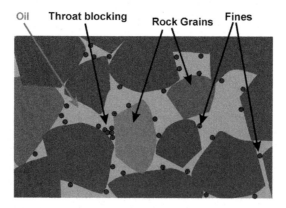

Figure 5.2 Schematic of fines in a grain structure.

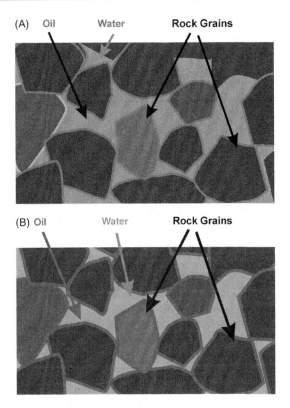

Figure 5.3 Water-wet (A) and oil-wet (B) rock with oil.

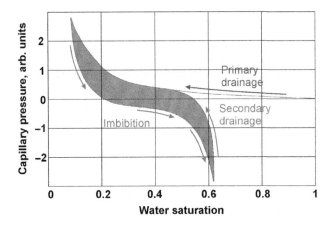

Figure 5.4 Dependence of capillary pressure on water saturation.

case the rock is water wet and initially is fully saturated with water ($S_w = 1$). At the initial drainage very small pressure is needed to extract approximately half of the connate water. As the water is drained further and the saturation goes below 0.3 the capillary pressure starts to rise sharply. One needs to apply significant external pressure to counterbalance the capillary pressure to continue drainage. Sharp capillary pressure rise means that more and more external pressure is needed to extract smaller and smaller amounts of water. At certain point one reaches the pressure at which drainage becomes unsustainable for various reasons. This is unreducible water saturation. If at this point the rock is brought in contact with water then water will be drawn into the rock − it is an imbibition process. However, in the case presented, the saturation would only reach the maximum at 0.2.

Remember, at this point, there is no external pressure to push water into the rock at this point. If we apply now external pressure then the external pressure will counterbalance the capillary pressure and the imbibition will continue. At first the imbibition continues relatively easily but then, after approximately 0.55 saturation value, the pressure needs to be increased quite sharply to push in smaller and smaller amounts of water. In the example provided we cannot get water saturation above approximately 0.65. The secondary drainage will create a hysteresis loop for water flow through the rock at a variable water saturations.

The explanation of the observed behavior is relatively simple. As saturation decrease only smaller and smaller pores would contain wetting phase and one needs more and more force to counterbalance the capillary force and extract smaller and smaller amounts of liquid.

The same is perfectly applicable to oil saturation in a rock. The behavior is more complicated for the multiphase system, but, generally, the behavior is similar. One only needs to remember that much also depends on the saturation history and the speed of saturation changes.

It can be seen that in the dynamic situation it is impossible to have full saturation and it is impossible to recover all liquid from the rock. There is always residual saturation − there is always oil and connate water left in the reservoir. Up to the point we can reduce residual oil saturation by either changing wettability or by decreasing interfacial tension but there is always oil left in the reservoir beyond economical recovery.

Oil extraction is a dynamic situation − liquids should flow though the interconnected pores and individual phase saturation is usually below 100%. In water-wet rock some pressure gradient needs to be applied for water to flow into the bigger pores, while making oil to flow into bigger pores requires some pressure in oil wet formation.

Partial permeability depends on the saturation too. In a simple model system water partial permeability k_{rw} and oil partial permeability k_{ro} dependences for a water-wet rock can be as shown on Fig. 5.5. It is possibly to say in general that the permeability for a phase will be at the highest at the highest phase saturation and that the permeabilities would be reduced to zero for the unreducible saturation.

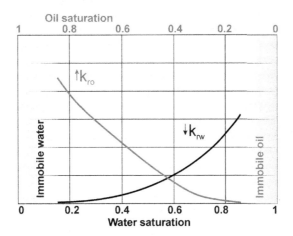

Figure 5.5 Dependencies of partial permeabilities on saturation for a water-wet rock.

Reservoir rock in general have very complex permeabilities for oil and water. The permeability can and will change from area to area for a whole number of reasons. Only laboratory experiments on multiple cores at the reservoir conditions allow to develop an accurate understanding of the reservoir processes and with certainty maximize oil extraction.

Introduction to well technology

6

Abstract

At present, both onshore and offshore, at all stages from the pursuit for hydrocarbons and exploration of hydrocarbon deposits up to the industrial development of oil and gas fields, along with geological, geophysical, geochemical and other studies, well drilling is employed. The wells are used for the access to the oil (gas) reservoirs, the assessment of hydrocarbon reserves, the development of deposits and the development of explored hydrocarbons. Oil extraction can also be done by use of excavated wells and a mines.

Chapter Outline

At present, both onshore and offshore, at all stages from the pursuit for hydrocarbons and exploration of hydrocarbon deposits up to the industrial development of oil and gas fields, along with geological, geophysical, geochemical and other studies, well drilling is employed. The wells are used for the access to the oil (gas) reservoirs, the assessment of hydrocarbon reserves, the development of deposits and the development of explored hydrocarbons. Oil extraction can also be done by use of excavated wells and a mines.

It is believed that the first in the world (July 4, 1847) industrially drilled oil well was done at the Absheron peninsula of the Bibi-Heybat field in the Republic of Azerbaijan, where an industrial oil inflow was obtained from a depth of 21 m. The first oil well in the United States was drilled in 1859 in the state of Pennsylvania by Edwin Drake, who produced oil from a depth of 25 m.

The French engineer Fowell invented and applied in 1848 the method of continuous wells flushing (removal of rock cuttings by a circulating stream of water) during drilling.

Washing circulating fluid flow for the first time was used for the first time in the United States in 1901. Next important step was taken with was an introduction of diesel and gasoline engines into the drilling practice. All this created global and prosperous oil industry.

Primer on Enhanced Oil Recovery. DOI: https://doi.org/10.1016/B978-0-12-817632-0.00006-2

The wells are drilled on land and at sea using drilling rigs that use conventional rotary drilling technology. The drill bit is connected to either drill pipes, which are joined using threaded box-and-socket joints, or to the flexible continuous pipes wound on a drum of pipe (up to 5000 m and more in lengths) – the so-called "coiled tubing technology".

Well drilling is a complex process consisting of the following basic processes:

- deepening of wells in rock destruction process by drilling tools;
- removal of drilled rock from the well;
- borehole fastening during casing drilling;
- undertaking a complex of geological and geophysical studies with the view to uncover rock properties and identify oil productive horizon(s) (strata);
- descent to the design depth and cementing of the last (operational) column.

When drilling for oil and gas, the rock is destroyed by drill bits, and the bottom hole is usually cleaned of the drilled rock with a stream of continuously circulating washing fluid (drilling mud). In some rare cases the bottom hole debris is blown with gaseous working agent.

A well is a cylindrical mine workings, with a diameter that is many times smaller than the total length of its trunk (depth). A well shape and arrangement excludes access to it by a person. The beginning of the well is called the mouth, the cylindrical surface – the wall or the barrel, the bottom – the bottom. The distance from the mouth to the bottom along the axis of the wellbore determines the length of the well, and the projection of the axis on the vertical – its depth.

The altitude of the well (absolute height) of the wellhead is the distance from the point of the earth's surface to the level of the surface adopted in the geodetic network as the original.

The maximum initial diameter of oil and gas wells usually does not exceed 900 mm, and the final diameter is rarely exceeds 165 mm.

When drilling a well, the rock is destroyed over the entire area of the face (solid bottom) or along its peripheral part (ring face). With the later method of drilling in the center of the well remains an intact column of rock – a core, which is periodically extracted to the surface for an inspection and laboratory study. The depths of oil and gas wells vary from the first tens to several thousand meters.

To date, the deepest oil well Z-42 (with a depth of 12,700 m) has been drilled at the Chayvo field in the Russian Federation under the Sakhalin-1 project.

6.1   Well construction elements

During the drilling process, it often becomes necessary to strengthen the walls of the well for various reasons. For instance, while drilling through unstable rocks, to prevent fluids cross-flow, etc.

If drilling of the following section (named interval) of the wellbore without wall strengthening of the previous section becomes impossible, then these sections of the wellbore are called "intervals with incompatible drilling conditions".

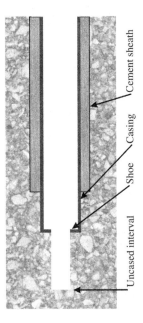

Figure 6.1 Main well elements.

The following terms are used in well construction:

casing – a string of pipes running into the well, having permanent or temporary adhesion to the well wall;
cement sheath – concrete, which fills the space between the borehole wall and the casing;
shoe – element of bottom casing;
cased interval of the wellbore – the interval along the axis of the well, in which its walls are lined with a casing;
open hole – the interval along the axis of the well, where there is no casing;
the output of the borehole from under the shoe – distance from the shoe to the bottom of the well.

The main well elements are shown on the Fig. 6.1. In many cases a well structure can be more complex

6.2 Well flushing and drilling fluids

Removal of drilled rock fragments from a well is one of the most important operations which is performed during drilling and it is carried out using drilling mud (drilling liquid mixture). The drilling mud should:

- clean the bottom hole of the drilled rock and bring the rock fragments to the surface;

- at the termination of washing to keep the drilled parts of the rock in suspension;
- prevent the wall collapse and penetration of gas, oil and water from the drilled formations into the well;
- create a mudcake of the borehole wall;
- cool the drill bit, the downhole motor and the drill string;
- lubricate moving parts;
- transfer energy to a turbo-drill;
- protect drilling equipment and drill string from corrosion and abrasive wear;
- be stable at elevated temperatures and inert to connate water and rock debris;
- be convenient for cooking, pumping and cleaning;
- allow prolonged cyclic use;
- be safe for the service personnel and environment.

One can see that it is long list of requirements to satisfy.

6.3 Well completion

For effective development and operation of the well in the process of drilling productive oil horizons, it is necessary to ensure the conditions of minimal disruption of the natural reservoir properties of oil-bearing or gas-bearing strata. In this case, productive formations after their drilling will be almost in the usual conditions and will be able to give out the maximum possible amounts of hydrocarbons during operation.

In the process of drilling wells to productive strata and putting the well into operation (production), the sequence of work is performed. It is usually referred as well completion. The well completion includes:

- initial reservoir assessment;
- selection of wellbore design;
- choice of wellhead equipment;
- provision for production string connection with the oil containing strata (secondary openings to productive strata by perforation);
- demonstration of oil flow at the ground level and well commissioning.

The well production, its operation profitability, the well production lifetime to the great extend depend on the results of the above works.

Also very important is the well drilling termination scheme (lower completion), which can be: an open hole, a filter with slit-like openings, an interval with partitions packers for strata isolation, or a lowered and then perforated casing. The choice of termination method can have a significant impact on the operation of the well and, therefore, drilling, completion, production engineers and geologists should evaluate various factors that influence the efficiency of the well operation.

Well classification can be carried out according to its intended purpose, according to its profile, according to operational and economic criteria. The classification is carried out in connection with the complex considerations of geological and economic factors. The classification adheres to the well-developed terminology for oil and gas reserves.

The well classification used in the United States is an improved classification which was proposed by F. Lahee in 1944. In accordance with the above classification all wells are divided into two main categories: exploratory and field wells.

Exploration wells are subdivided into five groups and are linked to the production and metering system for exploration during drilling.

The classification allows to assess the degree of economic risk in the production of exploration and gives a deeper content of the classification of reserves discovered at various stages of the exploration.

- **new-field wildcat** is a well located on a structural or non-structural perspective trap, within which no drilling operations were carried out or hydrocarbons have been produced. As a rule, these wells are drilled in areas with unexplored or poorly studied local geological conditions. Here the degree of risk taken by the entrepreneur in connection with the desire to test an oil trap, the productivity of which has not yet been proved by drilling, is high.
- **New-pool wildcat** is a well drilled to explore a new reservoir on a structural or non-structural trap outside the known boundaries of the productive area where oil or gas is already being produced. In some areas where local geological conditions have been studied not so well, such exploration wells are sometimes referred to as "close wildcat". The distance of the location of such a well from the nearest productive area usually does not exceed 3.2 km.
- **dipper pool (pay) test** − a well located within the boundaries of the productive area of a reservoir or reservoir that is already partially or fully developed. The purpose of the well is the exploration of the deep strata which lies below the reservoirs under development
- **shallower pool (pay) test** − a well being drilled in order to detect a new, not yet tested reservoir, the presence of which according to previously drilled wells can be assumed in the formations located above the developed or already developed reservoirs.
- **outpost or extension test** − wells are created for the purpose of prospecting a partially developed reservoir, usually at a double (sometimes somewhat longer) distance compared to the distance between production wells on the developed part of the reservoir.

  In the contemporary well classification, according to the data of T. N. Murray (1988), there can be three more groups of wells distinguished:
- **stratigraphic test** − a stratigraphic appraisal well is drilled to obtain specific geological information that can detect oil or gas deposits. Such wells are intended only for core sampling and (or) conducting any kind of well surveys. The wells are drilled without testing for productivity.
- **service well** − an injection well is drilled in the field under development in order to maintain reservoir pressure by pumping gas, water, air, steam, as well as for discharging reservoir water, obtaining water for injection and as an observation well.
- **old well drilled deeper** − a well drilled below its previous bottom hole depth, which may or may not lead to the discovery of industrial accumulations of hydrocarbons.

Field wells are production well which are drilled within an area of proven oil and gas formation to the depth of the productive horizon. The category may include not only the wells with which oil and gas is produced, but also wells for reserve estimation, injection wells and observation wells.

According to the results of production testing, exploration and production wells are divided into successful and unsuccessful (dry).

The spatial position allows to distinguish between vertical, inclined, and horizontal wells.

A vertical is a well whose axis deviation from the vertical passing through its mouth is within some prescribed limits. Historically, this is the oldest type of wells. Inclined well is deliberately drilled along a predetermined inclined to vertical trajectory. A horizontal well is an inclined well with the final interval deviating from vertical by more than 80 degrees. In many cases inclined and horizontal wells are referred to as deviated wells

Deviated wells are more expensive to drill as compared to vertical wells but if planned and implemented well they allow much better oil recovery from an oil field.

Horizontal wells are effectively used in the following cases:

- in fractured reservoirs, the location of a horizontal wellbore allows crossing fractures in the reservoir for effective drainage of the reservoir
- in reservoirs with the risk of water and gas breakthroughs, horizontal wells were used to minimize water content on the pumped to the surface liquids
- to increase oil recovery, especially in combination with the use of secondary and tertiary methods of stimulation.

6.4   Multilateral wells

It was justified long time ago that oil production from an oil horizon can be increased by the bigger diameter borehole well. It is reported that A. Grigoryan in 1949 in the former USSR had developed this idea further and drilled first well with the horizontal branching. While the horizon penetration increased only 5 times the oil production increased almost 20 fold.

Multilateral wells are wells that have branching in the part of the main bore in the form of two or more long horizontal, inclined or wave-shaped trunks. Oil production is carried out in the main trunk, and the branching serve as additional drainage channels through which oil comes from distant parts of the reservoir. The branches can access oil in highly productive cracks or lenses. At the same time branches can reach zones not affected by the previous reservoir development and stagnant zones of the reservoir.

The choice of the form of branching, the radius of curvature of boreholes depends on the geological, geophysical and lithological characteristics of productive reservoirs, reservoir pressure, natural reservoir conditions, methods of maintaining reservoir pressure, etc.

Currently, many different forms of branching and stem profiles of multilateral wells have been developed, differing from each other in the number of branches, their shape and length.

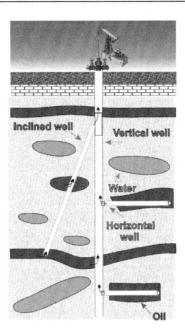

Figure 6.2 Multilateral well construction.

An example of possible multilateral well arrangement is presented on Fig. 6.2. The vertical well is combined with an inclined well and two horizontal wells. This construction allows to drain multiple oil reservoirs from a single surface pad. The combination of wells can be infinite and careful assessment of all available data and production modeling is needed for production optimization.

Oil recovery stages and methods

7

Abstract

The recovery of hydrocarbons historically was done through only two main stages: primary and secondary recovery. Primary extraction is carried out at the expense of the natural energy of the reservoir, while secondary extraction of hydrocarbons is carried out at the expense of the energy introduced into the reservoir from the outside, by injecting gas and/or fluids, to maintain or increase the initial energy in the reservoir.

Enhanced Oil Recovery comes after the above and aims to increase oil retrieval from the oil reservoir.

chapter outline

The recovery of hydrocarbons develops through two main stages: primary and secondary recovery. Primary extraction is carried out at the expense of the natural energy of the reservoir, while secondary extraction of hydrocarbons is carried out at the expense of the energy introduced into the reservoir from the outside, by injecting gas and/or fluids, to maintain or increase the initial energy in the reservoir.

7.1 Primary recovery

During primary extraction, the energy to move oil through reservoir into the production well is obtained by: gravity forces, expanding rock and liquid, releasing and expanding gas dissolved in oil while reducing reservoir pressure (depletion drive), expanding the gas cap or active aquifer, or a combination of these factors (Fig. 7.1).

Primer on Enhanced Oil Recovery. DOI: https://doi.org/10.1016/B978-0-12-817632-0.00007-4

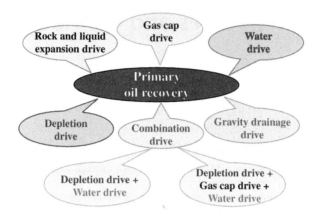

Figure 7.1 Driving processes for the primary oil recovery stage.

Figure 7.2 Driving processes for the secondary oil recovery stage.

7.2 Secondary recovery

The secondary recovery of hydrocarbons involves the introduction of energy from the outside into the reservoir through injection wells and the extraction of oil and gas from extraction wells. Typically, secondary recovery of hydrocarbons involves immiscible displacement by injecting water, gas or water-gas mixtures into injection wells (see Fig. 7.2). However, the most common fluid introduced by the reservoir to maintain its energy is water due to its availability and low cost.

As a rule, the introduction of secondary recovery in the early stages of reservoir development even before the natural reservoir energy is depleted leads to a higher oil recovery coefficient compared to the oil recovery coefficient obtained due to the action of natural mechanisms during the primary oil recovery.

7.2.1 Waterflooding

Waterflooding is one of the major oil production techniques. It is estimated that almost half of all produced oil is produced by use of waterflooding. Waterflooding is carried out by pumping water into a series of injection wells and hydrocarbons production through the production wells. Flooding, in general, is carried out to achieve any of the following goals, or combinations thereof:

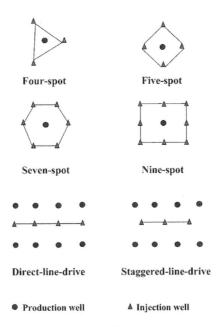

Figure 7.3 Some schemes for areal waterflooding.

- reservoir pressure maintenance;
- disposal of connate water after separation from hydrocarbons;
- creation of a water-pressure regime for displacing hydrocarbons from injection wells to producing wells.

Water injection is usually carried out by various waterflood schemes, such as areal (pattern flooding), contouring (peripheral flooding) and peripheral flooding (crested flooding). Field flooding is used in small depth but large area formations. Fig. 7.3 shows the possible wells arrangements of wells in the areal flooding. Economic factors, and, in some cases, ground topology, are the main criteria for choosing a specific well location geometry. The factors to consider include the cost of drilling new wells and the cost of transforming some production wells to the injection ones. Most often, straight-line (direct line drive) and alternate (staggered line drive) rows of injection wells are used (see Fig. 7.3). This is done on the basis of the minimal investment. However, if it is necessary to increase the injection rate, a seven-point or a nine-point flooding schemes are usually used (see Fig. 7.3).

In contour flooding implementation scheme, the injection wells are located along the perimeter of the reservoir. Fig. 7.4 shows two cases of contour flooding in the presence of underlying aquifers. In the anticlinal reservoir shown in Fig. 7.4A, injection wells are placed so that water enters the aquifer, or in the area near the water-oil contact, displacing oil into production wells located in the upper part of the reservoir. For the monoclinal reservoir shown in Fig. 7.4B, the injection wells are located below the water-oil contact to take advantage of gravity.

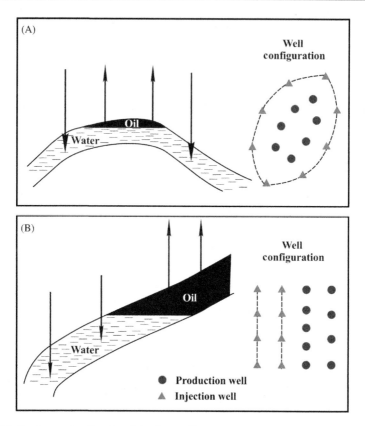

Figure 7.4 Contour water flooding injection well arrangements.

In reservoirs with an acute incidence angle, injection wells are positioned in the dome for efficient displacement of oil. This system is known as arch flooding.

In all cases, the areal configuration of the injection wells and the distance between the wells depend on several factors, which include rock and fluid characteristics, reservoir heterogeneity, optimal injection pressure, planned development period and economy.

The use of water flooding can increase the oil recovery ratio by 10−30%. The criteria for the applicability of waterflooding are shown in Fig. 7.5.

7.2.2   Gas injection

The immiscible gas injection is carried out to maintain reservoir pressure, slow down the production decline rate in the natural regimes of the reservoir and, sometimes, to support the gravity regime. An immiscible gas is usually injected with in alternating pattern with water. For immiscible displacement, associated petroleum gas, nitrogen or flue gases are used. The gas injected into the well behaves in the same way as the gas in gas cap drive mode: the gas expands acting as a compressed

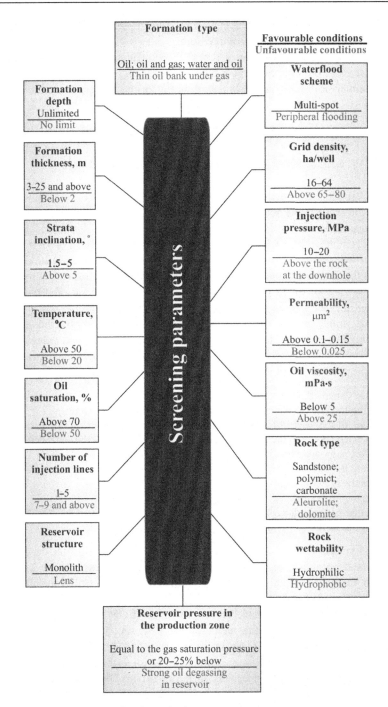

Figure 7.5 Waterflooding application criteria.

spring, displacing oil to the producing wells. The implementation of gas injection requires the use of high pressure compressors.

Injection of immiscible gases is inferior to the efficiency of water flooding, but in some cases it is the only possible method of secondary oil recovery. This is especially evident in low permeability reservoirs and in the reservoirs containing high amounts of swelling clays.

7.3 Why enhanced oil recovery (EOR) methods

Secondary oil recovery methods increase volumes of recovered oil. Nevertheless, despite the increase in oil recovery rate as a result of the use of secondary oil recovery methods, in particular the most effective oil flooding, a significant amount of residual oil remains in the reservoir. According to the accumulated around the globe work data for reservoirs: with low permeability (tight oil reservoirs) or containing heavy oils, the final oil recovery when using primary and secondary oil recovery is 5−10%; 10−25% of oil in place is recovered in the drive mode of dissolved gas; 25−40% oil in place is recovered at partially water-pumping mode, gas injection or gravitational mode; more, 40−55% oil in place, is recovered by the ubiquitous waterflooding. To increase the oil recovery rate above the indicated values, tertiary, so named Enhanced oil recovery (EOR) Methods of hydrocarbon extraction are used. The purpose of the EOR use is to increase the final oil recovery by:

- Increase in sweep efficiency due to:
    - reducing the ratio of the mobility of the injected and displaced fluids,
    - blocking of the washed highly permeable water-saturated zones and the re-direction of the injected fluid into the low-permeable oil-saturated zones of the reservoir.
- Surface forces modification in the reservoir due to:
    - reducing the interfacial surface tension between the oil and the displacing fluid,
    - reduce the effect of capillary forces,
    - changes in reservoir rock wettability,
    - disjoining pressure changes.
- Combinations of the above processes.

Ternary extraction processes of hydrocarbons are carried out after the processes of primary and secondary extraction have already been executed. At the same time, fluids other than those used in secondary recovery (ordinary water and immiscible gas) are introduced into the formation (see Fig. 7.6). As the above listed set of methods, EOR process is a broader concept because EOR methods can be applied at any stage of field development, including primary, secondary and tertiary hydrocarbon recovery. Thus, an EOR can be implemented as a tertiary process if it follows a flooding or injection of immiscible gas, but it can be a secondary process if it directly follows the primary recovery of hydrocarbons. Nevertheless, majority of EOR projects are implemented after water flooding. At this stage, it is important to establish the difference between EOR and IOR (improved oil recovery) to avoid misunderstandings. The term IOR refers to the use of any EOR operation or any

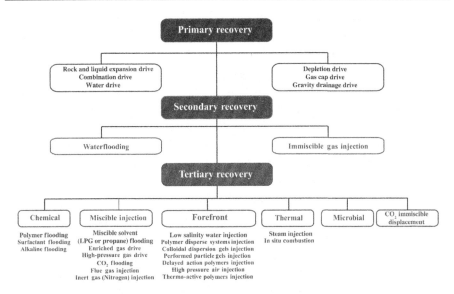

Figure 7.6 Oil recovery stages and applied extraction technologies.

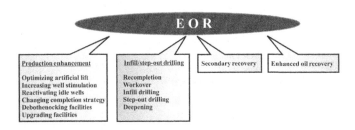

Figure 7.7 Enhanced oil recovery processes.

other advanced oil production technology that is implemented during any ongoing oil recovery process. Examples of IOR are any technology method leading to increase in oil production that is used during the primary, secondary or tertiary recovery of hydrocarbons. Other examples of IOR applications are: hydraulic fracturing, sedimentation prevention, acidizing, drilling additional wells to densify production well grid (infill drilling), and using horizontal wells (see Fig. 7.7).

Fig. 7.8 demonstrates current methods of enhanced oil recovery. Table 7.1 shows the underlying mechanisms (basic macroscopic processes) of enhanced oil recovery for the most used methods of enhanced oil recovery.

Chemical methods of enhanced oil recovery are characterized by the addition of chemical compounds to the injected water in order to modify the physicochemical properties of the fluid or interfacial tension. The later is more favorable for oil displacement. In polymer flooding, polyacrylamides or polysaccharides are used - this is a conceptually simple and inexpensive solution, and despite the fact that the technological efficiency of the method is not so high, it is used for commercial

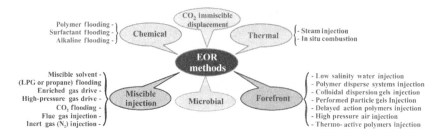

Figure 7.8 Enhanced oil recovery Methods.

Table 7.1 Underplaying mechanisms for increased oil recovery.

	Methods used	**Basic mechanism**
Chemical methods	Polymer flooding Surfactant flooding Alkaline flooding	Improvement of sweep efficiency; improvement of displacement efficiency
Miscible methods	Miscible solvent (LPG or propane) Flooding Enriched gas drive High-pressure gas drive CO_2 flooding Flue gas injection Inert gas (Nitrogen) injection	Improvement of displacement efficiency
Thermal methods	Steam injection In situ combustion	Improvement of both sweep efficiency and displacement efficiency

purposes. Waterflood with surfactants is a complex technology that requires detailed laboratory testing in the preparation of a field development project. This is a rather expensive technology and has been used only in a few large projects. Alkaline flooding is used only for the formations containing certain types of oil with a high acidic number.

Combined methods have the greatest potential for enhanced oil recovery for fields containing low viscosity oils. Among these methods, water flooding by mixing with CO_2 is probably the most promising.

Thermal methods are characterized by an injection into formation of thermal energy in order to reduce the oil viscosity. Low viscosity makes oil more mobile and this, in turn, increases the efficiency of oil displacement to the production wells. Steam injection, used since the 1960s, is the most advanced and used of thermal technologies for enhanced oil recovery. The steam injection efficiency can be estimated with less uncertainty than any other EOR. Internal combustion, which is currently in extensive development at the implementation stages, is usually applied

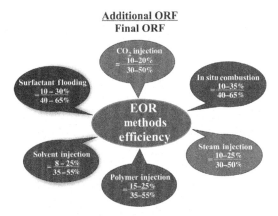

Figure 7.9 Additional Oil Recovery Factors (ORF) and Final ORF figures for the most used EOR methods.

to formations containing heavy oils. Many projects so far have been implemented, but only a few of them were commercially successful.

Fig. 7.9 shows the increase in oil recovery and final oil recovery figures, numbers (%) for the most used methods of enhanced oil recovery.

7.4 Sweep efficiency

The ultimate goal of enhanced oil recovery processes is to increase the overall oil displacement efficiency, which is determined by microscopic and macroscopic efficiency. Microscopic efficiency is provided on the scale of pores in the formation. It is determined by the interaction of the displacing fluid with the rock, interfacial forces and the fluid being displaced. For example, microscopic efficacy can be increased by lowering capillary forces, changes in wettability and wedging pressure.

Macroscopic or volumetric efficiency is observed on the scale of the entire formation. The efficiency of volumetric displacement is determined by the displacement efficiency both in the horizontal and in the vertical directions. It is also depends on how effectively the injected fluid displaces the oil to the producing wells. Figs. 7.10 and 7.11 show graphical representation of microscopic and macroscopic (horizontal and 3D sweeps).

The overall displacement efficiency of any oil displacement process can be increased by improving the mobility ratio of the displaced and displacing fluids. The mobility coefficient is defined as the ratio of the mobility of the displacing fluid (i.e. water) divided by the mobility of the displaced fluid.

Microscopic Sweep Efficiency

Figure 7.10 Microscopic sweep efficiency. All sizes are on a micron scale.

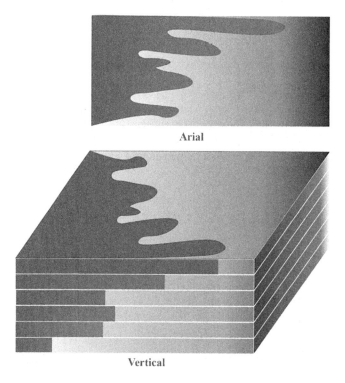

Arial

Vertical

Figure 7.11 Representation of macroscopic sweep efficiencies. Sizes are in tens of meters.

Mobility coefficient is defined as follows.

$$M = \frac{\text{Mobility}_1}{\text{Mobility}_2} = \frac{k_1/\mu_1}{k_2/\mu_2} = \frac{k_1\mu_2}{k_2\mu_1},$$

where 1 — displacing fluid; 2 — displaced fluid; k_1 and k_2 — permeabilities to 1 and 2, respectively; μ_1 and μ_2 — viscosity to 1 and 2, respectively.

Further reading

Baker, R., Dieva, R., Jobling, R., Lok, C., 2016. SPE-179536-MS, https://doi.org/10.2118/179536-MS.

Carcoana, A., 1992. Applied Enhanced Oil Recovery. Prentice Hall, Inc, New Jersey, USA.

Craft, B.C., Hawkins, M.F., 1991. Applied Petroleum Reservoir Engineering. Prentice-Hall, New Jersey.

Lyons, W.C., Plisga, G.J., 2005. Standard Handbook of Petroleum and Natural Gas Engineering. Gulf Professional Publishing, USA.

Romero-Zerón, L., 2012. Advances in enhanced oil recovery processes. In: Romero-Zerón, L. (Ed.) Introduction to Enhanced Oil Recovery (EOR) Processes and Bioremediation of Oil-Contaminated Sites, ISBN: 978-953-51−0629-6.

Satter, A., Iqbal, G., Buchwalter, J.L., 2008. Practical Enhanced Reservoir Engineering-Assisted With Simulation Software. Penn Well.

Thermal EOR

8

Abstract

Industrial oil extraction continues for more than century. As the result, majority of the exploited for long time reservoirs contain mostly difficult to extract oil. We can say that both types of oils — heavy and viscous, are dominating. Low mobility of oil in porous media significantly hardness effective extraction. At the same time, oil viscosity significantly defined the temperature, when the viscosity is significantly reduced as temperature rises. It is therefore economically fishable to rise temperature in the formation if initial oil viscosity is in the region of above 100 cp.

Chapter Outline

Industrial oil extraction continues for more than century. As the result, majority of the exploited for long time reservoirs contain mostly difficult to extract oil. We can say that both types of oils — heavy and viscous, are dominating. Low mobility of oil in porous media significantly hardness effective extraction. At the same time, oil viscosity significantly defined the temperature, when the viscosity is significantly reduced as temperature rises. It is therefore economically fishable to rise temperature in the formation if initial oil viscosity is in the region of above 100 cp.

Processes of thermal extraction rely on transferring thermal energy from different sources to rise the temperature in the formation. Higher temperature provides for lower viscosity and more effective movement of oil to the production well. It is

Primer on Enhanced Oil Recovery. DOI: https://doi.org/10.1016/B978-0-12-817632-0.00008-6

possible to say that thermal methods are most used techniques in EOR worldwide. Two methods are most widely employed — steam injection and internal combustion. We have to stress at this point that when we are talking about steam injection we are exclusively talking about water steam injection.

8.1 Steam injection

First experiments with steam stimulated oil extraction were conducted by Stovall at around 1930. On the modeled formation at around 98% Original-Oil-In-Place (OOIP) was extracted by the steam stimulation. This immediately has been followed by the industrial application to an oil field in Texas. In 1939 Lapuck has published first review of works on steam injection. Next wave of work in this direction started after Second World War when in 1951 Winkler conducted stem injection oil extraction in Leoprechting (Austria). Injected steam temperature (in the well on the ground level) was in the region of 200–220 °C. It was shown that the method was economically feasible.

Wide utilization of steam injection started at around 1960 both in the USA and USSR. It is still possible to say that the stem injection is most widely used EOR method for high viscosity oils.

Continues steam injection consists of steam injection into the well at around 80% steam saturation. This stem, after reaching oil, condenses and transfers the latent heat to the formation. Released thermal energy rises the formation temperature and lowers oil viscosity. At the same time at the elevated temperature light hydrocarbons separate from the oil in form of gas. This gas moves in front of propagation hot water — water steam front and increases oil mobility and, ultimately, oil displacement towards production wells.

Cyclic stem injection as also used, it is named *huff and puff method*. In consists of utilizing production wells for first injection of steam (huff phase), soaking period and production period (puff phase). During steam injection some volume of formation is heated by the injected steam. Soaking allow to utilize injected energy more economically. During both those phases oil viscosity is reduced and partial oil upgrading is reached. Then during the huff phase more mobile and partially upgraded oil is extracted.

8.1.1 Detailed description

Continuous steam (water vapor) injection is used in oil formations with low temperature and high oil viscosity. The injection usually proceeds through especially devoted steam injection wells. Water steam has high hidden energy of condensation (this is energy released during steam condensation into water liquid phase). Moreover some energy is released during steam itself cooling from injection to condensation temperature. As the result during cooling from, for example, 230 °C at the injection point to cooling down to formation temperature, let us say at around

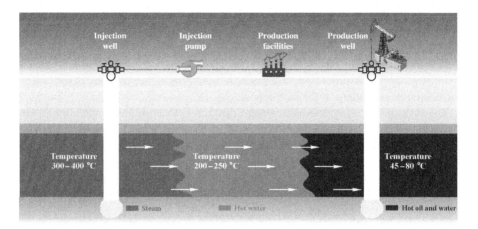

Figure 8.1 Steam injection into a formation.

60 °C, one kilogram of steam will realize more than 5000 kJ. This is considerable amount of thermal energy which will increase temperature of rock with oil. Oil viscosity will be reduced and oil will move more easily towards the production well. Moreover, the steam is very effective in removing oil from small cavities.

It is convenient and useful for process understanding to divide processes in the formation under steam influence into few zones. In oil containing strata three distinctive zones can be identified. Those zones are classified by the significant temperature differences and by the prevailing mobile phases.

On the Fig. 8.1 it is assumed that we inject steam at the temperature at around 400 °C (supercritical fluid) and the reservoir temperature is at around 45−80 °C.

Zone 1. Temperature is almost the same as temperature of injected fluid. Partial steam condensation starts at temperatures below 374 °C (critical temperature). In this zone we have oil distillation, partial extraction of oil light fractions and removal of those fractions into the next zone.

Zone 2. In this zone we have temperatures at which most of steam condenses into the hot water. Hot water proceeds through the formation and displaces the hot oil towards production well.

Zone 3. No water steam left. Temperature slowly drops over distance to the normal oil formation temperature. Oil mostly displaced by the water, which now has properties (temperature and salinity) close to connate water.

After the beginning of steam injection zones one and two grow and zone three shrinks. The injection process continues to the point when breakthrough occurs. Injection stops when water cut approaches approximately 80% and the process becomes not economically feasible anymore.

Many other processes develop during oil formation heating by the steam. We have already mentioned oil viscosity reduction and partial oil distillation (upgrading). We also have materials volume increase, phase penetration (mixing) changes, change of rock wettability by the reservoir fluids, modification of connate water salt content and water viscosity to mention few. The steam has much smaller density than the formation fluids. There is always vertical gravitational drift of steam and this effect needs to be taken into the account.

However, the main effects generally are the oil viscosity reduction and the oil distillation. All other processes have rather limited effects and in many cases can be considered as secondary at the planning stages.

The main goal during continues steam injection is to raise the temperature in the formation, especially in the vicinity of production wells, reduce viscosity, increase pressure and enhance oil flow. It is evident that all this leads to increase in oil production.

Basic processes during cyclic steam injection (huff and puff) are the same as during continuous steam injection. One needs to remember that in this case we use the production wells for the steam injection, which should be able to withstand injection pressure. The difference in the processes is then in the phenomena development direction. Injection of the steam is followed by soaking. The soaking is an essentially waiting time so that the heat can penetrate the formation area around the well. During soaking the direction of heat and stem/water flow is opposite to the oil production stage flow direction. During initial steam injection stem on the first place penetrates into the big openings which have big penetrability. During soaking period there is slow re-distribution of steam and hot water into smaller and smaller capillaries. As the result small capillaries will release oil into the big opening with high penetrability.

It is possible to say that during cyclic steam injection main effect in increase of oil recovery is provided by the rise of penetrability of oil in the areas around the well. Moreover, the well drainage zone is increasing, further growing oil recovery from the formation.

Development in ability to produce horizontal wells had allowed in 1980 in Canada to develop technique named Steam Assisted Gravity Drainage (SAGD). Over time this method became the industry standard in production of very viscous (bitumen type) oil. In implementation of this method two paired horizontal wells are drilled. Steam injection well is located vertically above production well. Again, injected steam heats heavy oil and reduces oil viscosity. Mobile oil under gravitation force sinks into the production well and is pumped to the surface.

8.1.2   Application criteria

One of the potential major difficulties for steam injection method is access to high quality water, e.g. water with low salinity (mineral content) and low organic impurities content. The impurity content should be below 5 μg/l. Moreover, presence of Mg and Ca cations should be particularly avoided. Otherwise production of high value steam (steam quality above 80% and thermal capacity 5 MJ/kg) is impossible. Due to this limitation, water needs to be chemically purified with expensive chemical reagents. Cost of the chemicals can be as high as one third of steam production coast.

If the oil containing strata is relatively thin then this leads to the situation when ratio of strata volume to the surface becomes relatively low and this ultimately leads to high thermal energy loss into the surrounding oil-less formation. This ultimately means that the maximum temperature in the reservoir will be lower and temperature uniformity will be compromised. The same can happens if the density of production/injection well ratio is below one per hectare.

It can be reasonably expected that for the oil formation depth at around 900 m and initial oil viscosity 1000 mPa-s it is possible to extract around 55% OIP with steam injection against 15−18% OIP by water flooding.

In the favorable conditions for one additional oil barrel extracted with the steam injection one will need at around 2.5−3.5 barrels of water converted into the steam and injected into the formation.

Oil is usually burned in the steam generators. It is realistic to expect to spend one barrel of oil to generate 2.5−3.5 barrels of water converted into the steam. The final balance means that for each barrel of oil burned in the steam generator one expects to extract 4−5 barrels of oil. This is sensible economy expectation for steam injection project.

In the case of loose sand oil containing formation steam injection process activates sand mobility and a lot of sand is pumped to the surface. This makes overall process more difficult and lowers profit margin.

One needs to bear in mind that the stem and hot water after steam condensation have much lover viscosity than the connate or injected water at the formation temperature. This leads to much more unstable water/oil displacement front and earlier breakthrough.

For deep formations, at or above 900 m, heat loss from the steam in the injection well becomes very significant, as despite efforts to insulate the injection well from the surrounding rock it is expected that the heat losses from the steam will be at around 4% per every 100 m of depth. This loss can be reduced by good insulation but still remains very substantial at high depth. One way to deal with this problem is to increase injection speed as this reduces heat losses.

Instead of continuous steam injection it is possible to do cyclic injection. This reduces cost of injection process but reduces oil extraction to 10−30%. Porosity for cyclic steam injection should be above 25% and thickness of formation above 6 m at the depth 900 m. It was shown that variable cyclic periods can gain additional 10% on the numbers mentioned above.

Huff and puff method requires the lowest initial investment outlay but only produces 3−4% of OIP.

For successful SAGD implementation the formation should be formed by not fully densified sand-type with high vertical permeability.

It is shown on the Fig. 8.2 major criteria for steam injection implementation.

8.1.3 Implementation methods

There are two major implementation methods. In the traditional one (see Fig. 8.3) 80% saturated stem is produced in a steam generator (Fig. 8.4). This method

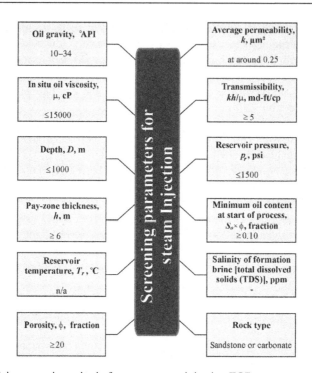

Figure 8.2 Major screening criteria for water steam injection EOR.

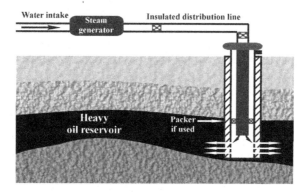

Figure 8.3 Schematic of steam prepared on the ground level and injected through well thermally insulated pipes into well thermally insulated injection well.

produces significant air pollution and is generally associated with significant heat losses, even despite thermal insulation of steam

Both, distribution network and injection wells lead to the heat loss. In some cases the heat loss can be as high as 1/3rd. On the Fig. 8.4 there is an example of steam generator unit closely placed to the injection well in order to avoid heat losses in the distribution network.

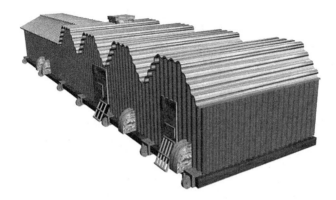

Figure 8.4 Steam generator unit.

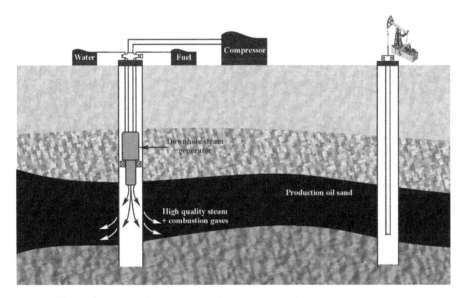

Figure 8.5 Utilization of downhole steam generator.

The second method involved utilization of downhole steam generator. As can be seen on the Fig. 8.5 an oil burner is inserted into the well while fuel, air and water are pumped from the surface. All three essential components for steam generation are injected at high pressure. Almost all steam produced keeps the thermal energy while reaching the oil bearing strata. Additional benefit is produced by carbon dioxide and nitrogen. The carbon dioxide is produced by the oil burning; the nitrogen is just leftovers from the air after all oxygen is consumed in burning process. All burning products are injected into the oil formation and do not damage the environment.

During cyclic steam injection (see Fig. 8.6) the process is divided onto three stages. During the first stage steam is injected into a well. This stage can last up to

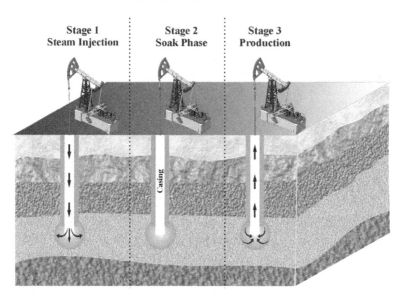

Figure 8.6 Cyclic steam injection stages.

a month and usually between 30 and 100 tons of steam is used for each vertical meter of formation thickness. The exact steam volume is defined by the oil viscosity and temperature in the formation. The volume goes up as the viscosity is higher and initial temperature in the formation is lower.

Soaking phase is stage 2. The injection ceases, the well is sealed and all activity is on suspension. This allows heat to spread and to release oil from the small pores. At the end of the stage all steam condenses into water. The influenced area contains hot water and oil bubble. Elevated temperature reduces the oil viscosity. Oil also leaves small pores. This makes more oil accessible for the extraction.

During stage 3 the well is switched into production. This stage continues until the flow of oil is economically feasible. It is usually lasts 8−12 weeks.

The all cycle can be repeated few times. Soaking stage is getting longer and longer at each cycle and amount of extracted during third stage falls from cycle to cycle.

8.1.3.1   Steam assisted gravity drainage (SAGD)

For application of SAGD two horizontal wells are drilled one on top another with vertical separation at around 5 m, as shown on Fig. 8.7.

Calculations show that length of horizontal wells can be up to 1000 m. Usually the process starts with steam injection into both wells. Then deeper well switches to oil production mode. Heat from the top well slowly penetrates through formation, rises formation temperature and the oil viscosity drops. Earth gravitation force drives oil deeper and, ultimately, into the production well. Condensed water is

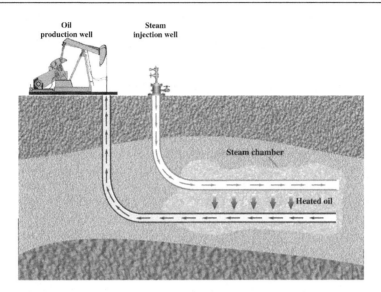

Figure 8.7 Well arrangement and schematic of oil extraction process during SAGD.

pumped with the oil. Continues contact of steam with oil provides for low parasitic heat loss and high efficiency of the process.

8.1.4   Implemented projects

Steam injection has become widespread in countries with high viscosity oil deposits, mainly in Canada, the former Soviet Union, the United States and Venezuela. Several projects have also been implemented in Brazil and China. It should be noted that steam-gravity drainage (SAGD) technology is used for bitumen mining, which is implemented at the Alberta fields.

Injection of steam into the reservoir to increase the recovery rate of the reservoir began about sixty years ago. In the past fifty years, steam injection has been used at the Mene Grande and Tia Juana fields in Venezuela and Yorba Linda and Kern River in California. Later, continuous steam injection was used at the Crud E fields in Trinidad, Schoenebeck in the Netherlands and Alto do Rodrigues in Brazil.

In the former USSR, steam injection into reservoirs began in the 60s in Azerbaijan to the Khorasany area of the Balakhani-Sabunchu-Ramany field, the Binagadi-Kirmaki area, the Binagadi-North of the Binagadi field, the Pirallahy field; Krasnodar region and Ukraine. A technology was used to displace oil by injecting a thermal rim, subsequently moved by water injection, at the Okha (Sakhalin) field. The technology showed high efficiency and was also implemented at the Yaregskoye and Kenkiyakskoye fields. Since 1982, steam injection began at the Karazhanbas deposit (Kazakhstan). The steam was pumped into 27 injection wells. The volume of steam injection exceeded 400 thousand tons/year. It is estimated that the additional oil production is at around 150 thousand tons/year.

At the pilot site of the Usinskoye field (Russia), for the first time in world practice, equipment and technology were implemented to work on a deep carbonate reservoir (steam temperature 320−330 °C, formation depth at 1400 m).

Modifications of the steam injection method, in particular, the combination of continuous and cyclic steam injection into reservoirs are used. It was applied at the fields Kern River, San Adro, Weig Wolf (California, USA). The depth of the deposits is 200−600 m. The thickness of the reservoir is 25−70 m, the viscosity of the oil is more than 3000 MPa s. Since the 1960s, a combination of continuous and cyclic steam injection into more than 2500 wells per year has been used at these fields. As a result of the introduction of recoverable oil reserves reached 35−37% of the geological.

The gravitational drainage technology with steam injection (SAGD) is being introduced in bitumen and super heavy oil fields, especially in Canada and Venezuela.

8.2 In-situ combustion (ISC)

In situ combustion is a process involving in a reservoir oil burning when oxygen containing gas is injected into the oil containing formation. Oxygen reacts with hydrocarbons, energy is released and the formation temperature rises. Continues gas injection allows to create self-sustaining moving burning zone with temperature 200−650 °C. The ignition can b provided by the means of electrical discharges or gas burners. In some cases oil in the formation will ignite itself in oxygen presence in the formation. In many cases just air pumped into the formation. The other component of burning, as the fuel, remains of heavy oil (after upgrading zone passes) will be used. Burning thermal energy increases formation temperature, this reduces oil viscosity, partially distills and evaporates oil. Water steam, hot water and gases misplace the oil towards production wells (Fig. 8.8). ISC is possible only if the burning process generates enough heat (apart the situation when heat is supplied continuously from the outside) to sustain itself.

Two main types of ISC are known − forward combustion and reverse combustion. In the first case burning zone moves with the direction of the injection gas (oxidizing media). In the latter case burning front moves in the opposite to the oxidizer direction.

The ISC method was first proposed in the former USSR by Scheiman and Dubrovai. In 1934 they first applied it at Schirwan oil field in Maikop. The ISC was later very widely used in the USSR.

From 1960 the method has been widely used in the USA and Venezuela. In 1964 the method was applied at Suplacu de Barcau (Romania) with the recovery efficiency 54%. In the last 20 years the technology is widely used in Canada and India.

8.2.1 Detailed ISC description

The ISC method relies on burning some part of oil in the formation, at the same time significant part of the oil is pushed towards production wells. In many cases

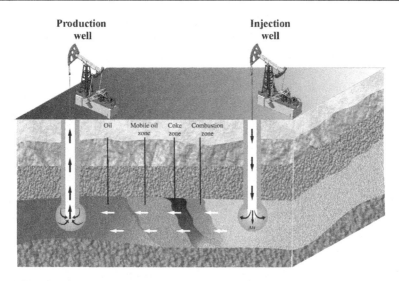

Figure 8.8 In situ combustion process.

the burning mostly proceeds in the relatively close vicinity of air injection well. It is necessary to have the temperature at around $65-95$ °C to start the burning. In order to rise the temperature to this level various methods such us electrical heaters, gas burners and various exothermic reactions. The burning first ignited in the relatively confined volume and burning products are extracted to the surface. As temperature rises to $200-650$ °C then the burning front can move through the formation itself in the presence of sufficient amount of oxygen. Approximately $6-25\%$ of oil would burn. This amount depends on the formation density, oil viscosity and permeability. The amount of burned oil is proportional of sorts to the first two and reciprocal to the last.

Forward combustion can be implemented by injection of air (so named dry burning) and air/water (wet burning). Air and water can be either injected together or sequentially. Single injection well is used in this case.

Reverse combustion is done by utilization of two injection wells. One well is used to ignite the burning. The second well is used to inject air, which moves towards the burning zone. The oil is then moves towards the first well which needs to be switched to the oil production. The reverse combustion is used for very heavy, immobile oil, where forward combustion does not work.

8.2.1.1 Dry forward combustion

In the forward combustion the burning front acts as a piston, which pushes oil in front of it towards the production well. Oil is partially distilled (upgraded). Light oil fractions are pushed by the burning front. The heavy, slowly moving or immobile fractions, are left behind and are consumed by the burning.

During dry burning we can identify four zones, as on the Fig. 8.9

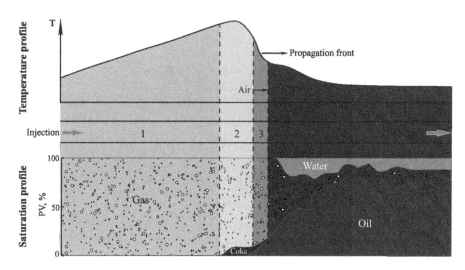

Figure 8.9 Zoning for dry forward burning.

Zone 1 is fully burned volume. It is area between injection well and burning zone. In an ideal situation it is the formation zone completely dry (oil free) rock. Injected air is heated in this zone. The temperature rises as we proceed from injection well towards the burning zone.

Zone 2 is burning zine. The burning processes with consumption of oxygen, carbohydrates and coke. Temperature in this zone depends on the formation geology and reactants concentration/composition.

Zone 3 is coke formation volume. This zone significantly alters hydrocarbon content and composition. In this zone the lighter reservoir hydrocarbons either evaporate (become gaseous) and all other mobile hydrocarbon species are pushed towards production well. Unmovable hydrocarbons are pyrolysed and have very low mobility.

Zone 4 is characterized by gradual temperature decrease by the distance from Zone 3. There are no chemical alterations to the hydrocarbon composition. The zone is filled by combustion products (oxides and water steam), gaseous hydrocarbons and high mobility liquid hydrocarbons. The following needs to be noticed:

1. On the border with zone 3 we have continuous evaporation of light oil fractions and partial condensation of water steam
2. In the further removed zone we have continuous and, at the zone edge, full water vapor condensation. Additional, compared to original connate, water creates so named wall-water bank. This additional water effectively displaces mobile oil from the pores. In the case of very immobile oil this water ingression can plug the formation.

8.2.1.1.1 Wet forward combustion

During dry forward temperature of the formation still rises in zone 1. This happens due to low heat capacity of air (injected oxidants). At around 80% of generated heat energy remains in zone 1, does not contribute to the oil extraction and is

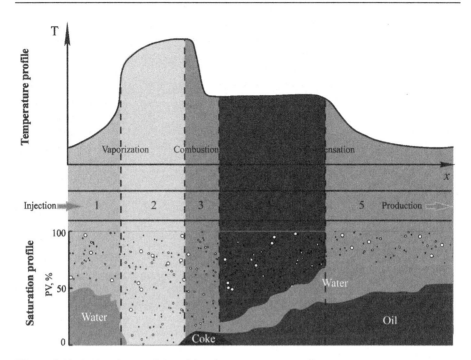

Figure 8.10 Zoning for wet forward burning.

eventually just lost. Additionally, almost 40% of costs are related to the air injection into the formation. All this increases recovery expenses. It is possible to reduce those problems if the heat conductivity of gases is increased. This can be done if one adds something to provide heat conductivity and/or transport. It is possible to use water. On those grounds technology of wet forward combustion has been developed. Water then either injected together with air or sequentially. Water evaporates in zone 1 and transfers heat beyond burning zone. It is possible then identify five zones, as on Fig. 8.10

Zone 1 does not contain almost any hydrocarbons. This zone is saturated with water and oxygen. As temperature is below water boiling point, the zone contains much of liquid water.

Zone 2 has all water in vapor form. The line of full water evaporation divides zone 1 from zone 2.

Zone 3 is burning zone. Burning consumes all hydrocarbons, including unmovable coke.

Zone 4 encompasses both evaporation and condensation. Temperature in this zone is close to water evaporation at the same time condensation of water from burning products is taking place. Light fractions of oil evaporate. On the border with zone 3 partial hydrolysis takes place.

Zone 5. Contains main water and oil volume. Temperature and pressure are close to the formation original. The mobile hydrocarbons are moving towards production well.

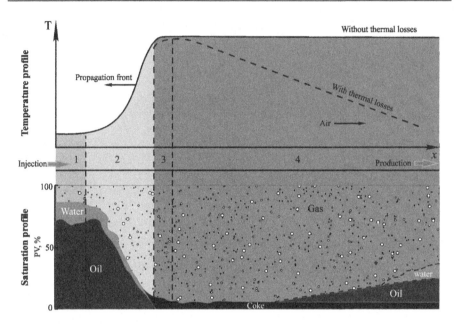

Figure 8.11 Zoning for reverse combustion.

8.2.1.1.2 Reverse combustions

Zone 1 near injection well is considered to be in as-formation conditions prior to air (oxidant) injection. However, if oil is easily oxidized, this can lead to slow increase of temperature following beginning of the injection (Fig. 8.11).

Zone 2 includes formation with steeply rising temperature due to heat transfer from burning zone. The heat is railed by hot gases and water vapor from burning zone. The zone is also significantly enriched by light oil fractions distilled from the burning zone. Oil light and mobile fractions are displaced. Some amount of coke is also created.

Zone 3 is burning zone. Temperature is at its maximum. Burning consumes all available oxygen. As the rule only most reactive hydrocarbons oxidize.

Zone 4 can be characterized by well separated hydrocarbon and water fractions. Due to thermal losses to the surrounding rock, the temperature in this zone decreases with the distance from burning zone. The decrease rate depends on water and light oil fraction condensation conditions. Produced during burning coke is immobile while light oil fractions and water are very mobile and can be easily extracted.

8.2.2 Screening criteria

In situ combustion is applicable to wide type of oil reservoirs (Fig. 8.12). In general terms ISC application conditions, e.g. application boundaries, are identical to steam injection. Oil properties, formation depth, possible amount of energy produced and ability to collect processed oil need to be carefully considered.

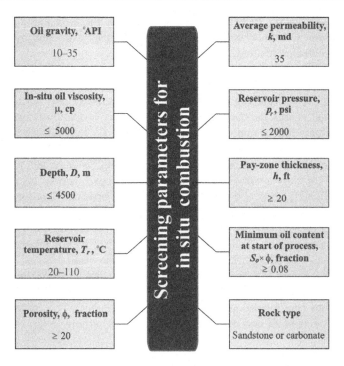

Figure 8.12 Screening parameters for in situ combustion.

First and probably the most important criteria is the formation depth. While it is widely accepted that the depth limit is at around 3500 m, the guidance depth is not the real limit. Mostly the challenge is related to the ability to inject air at the formation pressure and ability of crude oil to produce enough heat to sustain burning front propagation.

Historically it is accepted that the formation pressure should be below 140 atm. In reality this is not the limit. The challenge is always to be able to maintain injected air volume at the necessary pressure. There are application examples when the pressure was well above 140 atm.

The formation permeability is the third criteria. This is where steam injection differs from ISC. ISC can be used in the formations with lover permeability, as compared for stem injection. Challenge is to provide sufficient amount of injected air at the formation pressure. Amount of air is calculated on the basis of oil and formation properties. The permeability should be just enough to sustain air delivery to the sustained burning zone.

For the ISC application the formation can contain medium to heavy oil. The method can be used for oils with API between 35° and 10°. As the rule, operating in this range does not require additional fuel infection into the formation.

The existence of high API degree limit is only surprising at the first glance. In reality, if the oil is too light, almost all oil evaporates and coke does not form. Absence of coke then does not allow to sustain continuous burning. On the other side, if oil is too heavy then during forward burning the coke can start blocking steam dispersion. For this reason for heavy oils preferred extraction by application of reverse combustion method.

As it is clear from the above, the bitumen content is very important. It is accepted that the lover limit for the sulfur containing bitumen content is at around 20 wt%.

The oil deposit (oil bearing zone) thickness is very important. As the rule up to third of thermal energy can be lost to the surrounding formation if the oil deposit has thickness between 10 and 6 m. Below 6 m thickness heat loss can be too high for sustainable burning.

The best formation stratification arrangement for ISC is combination of oil containing stratas of clay layers. Clay strata is very effective thermal insulator, which reduces significantly vertical thermal losses. The clay provides then for higher temperature and higher sweep figures.

During ISC there is tendency of upward burning front movement. This is mostly gravity meditated process. If the oil containing strata is too thick (above 20 m) then sweep efficiency can be significantly reduced by this process.

In summary, before ISC application one needs to consider formation depth, formation thickness, oil gravity and oil composition. In the ideal case scenario laboratory trials with the formation cores need to be conducted in order to define necessary amounts of air and water, especially for wet burning.

8.2.3   The implementation technology

The process starts with heater insertion into a well as this is widely accepted step. Propagation of heated zone is assisted by injection of air, which spreads heat into the formation. Later some water is added to the injected air.

Burning ignites either spontaneously or by special devises lowered into the formation. Spontaneous ignition is possible in the cases when oil has high oxidation activity. As soon as oil starts to oxidase the reaction process releases thermal energy which increases temperature and stimulates speed of oxidation process further. At certain stage temperature rises to the level of burning.

Forced ignition is produced by special electrical or flame burners. In some cases easily oxidizing liquids (linseed oil) are injected into the proximity of injected air.

In the case of forced ignition the burners are lowered into a well on the steal cable, tope or tube. The implementation is shown on Fig. 8.13.

At the beginning of the process it is possible to increase temperature at the surrounding well formation by approximately 260° in 24 hours.

The most common method of initiating combustion of reservoir oil is carried out utilizing downhole electric heaters, as it is less complex compared to flame burners. For lowering electric heaters into the well reinforced electrical cable is used. Usually the electric power is supplied from the field network but it is possible to use mobile electric generators.

Oxygen in form of air needs to be supplied at the required high pressure and volume. Specialized compressor equipment is needed and has been developed. Field experience in the implementation of in situ combustion shows that to carry out the

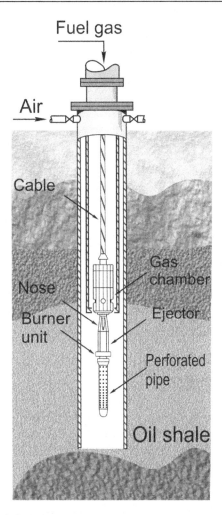

Figure 8.13 Deep gas-air heater.

process it is necessary to pump between 20 and 250 thousand m^3 of air per day. The air pressure should be in the region from 20 to 90 atm. The choice of compressor is carried out on the basis of the calculation of the required air volume and injection pressure. There are three types of compressors used: piston, centrifugal and axial types. Axial compressors are selected when the need for high air volume flow is required, while in case of a need to increase the outlet pressure, centrifugal compressors are installed.

The mouth of the ignition well is prepared as shown on Fig. 8.14 for the case of the downhole ignitor. In the case of ignition of reservoir oil with a submerged fired heater, the mouth of the ignition well is arranged as shown on Fig. 8.15. As can be seen that the implementation of the process of in situ combustion does not require too special wellhead reconstruction.

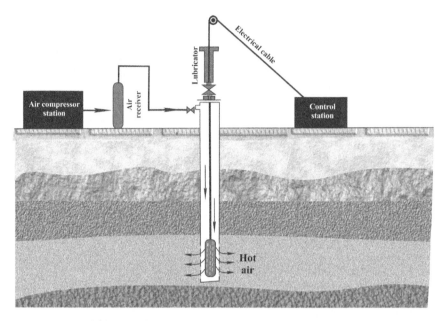

Figure 8.14 Ignition well arrangement for ignition by the downhole electrical ignitor.

8.3 Borehole design

The main requirement for the design of both injection (ignition, incendiary) and production wells is to prevent the rapid well material deterioration as a result of corrosion stimulated by high temperature. In this regard, the shanks of casing pipes must be made of alloys with special heat-resistant and oxidation-resistant properties.

The completion of ignition wells, where possible, should be carried out with an open casing, with a freely suspended tubing. This construction provides sufficient thermal insulation.

In production wells with poorly cemented rocks, where there is a danger of destruction of the bottomhole zone and significant sand removal. Exsessive sand extraction can be avoided by the installation of special anti-sand heat-resistant ceramic filters.

8.4 Well products separation

During the implementation of in situ combustion the burning products contain many highly active chemical compounds and cause extensive equipment corrosion. To minimize this the main part of gases is removed through the annular space of the wells. It should be mentioned that unpurified gaseous products of combustion, when released into the atmosphere, lead to significant environmental pollution. Safe disposal of those products need to be undertaken.

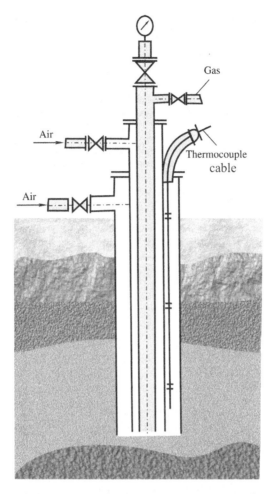

Figure 8.15 Ignition well arrangement for ignition by the submerged fired heater.

Of great importance is the produced gas composition monitoring. The monitoring allows to estimate the location of the combustion front and to regulate its propagation. In addition, the gas composition analysis makes it possible to assess the degree of possible air pollution during their release and actively implement safe disposal strategies. It is also important to monitor the reservoir fluid properties of in order to define the water content in the extracted product. It is highly recommended to monitor and log the oil acid number.

8.5 Implemented ISC projects

As was mentioned before, first ISC projects have been realised in the USSR at around 1940s. Well documented projects have been realised in the USA in 1960s.

Sloss Formation (Montana, USA) with light oil has been extensively developed before 1960s. The secondary oil extraction phase ended at around 1963 after full water flooding. In the next six years approximately 19% of oil remaining after water flooding has been extracted by implementation of wet ISC. This implementation has demonstrated effectiveness of ISC processes which was done after water flooding.

Dry ISC was implemented at Suplacu de Barcău (Hungary) in 1966. Some wells increased production in more than 10 times. Average air consumption was at around 2200 m^3/tons.

In the former USSR first project was realised in Azerbaijan in 1973. Production increased by 25% and the average air consumption was at around 1840 m^3/tons of produced oil.

It is estimated that in 1980s more than 50 ISC projects have been realised worldwide. It is known that in the USA additional oil production was at around 600 thousand tons/year, while in Hungary it was at around 430 thousand tons/year.

It is argued by Alvarado and Manrique that despite long history and proven commercial outcome majority of oil field operators do not really believe in OSC technology. This skepticism is based on high portion of unsuccessful projects and projects with unknown outcomes. At the same time it is also accepted that failed applications are due to ISC applications on formations which are not really suitable for the technique.

Further reading

http://www.tatar-inform.ru/news/2013/04/27/358128/.
Alvarado, V., Manrique, E., 2010. Enhanced Oil Recovery: Field Planning and Development Strategies. Gulf Professional Publishing.
Amelin, I.A., 1980. In situ combustion. M.: Nedra.
Baibakov, N.K., Garushev, A.R., 1981. Thermal methods of oil field development. M.: Nedra.
Burger, J., Sourieau, P., Combarnous, M., 1984. Recuperation assistee du petrole les methodes thermiques. Editions Technip, Paris.
Carcoana, A., 1992. Applied Enhanced Oil Recovery. Prentice Hall, Inc, New Jersey, USA.
De Haan, H.J., Van Lookeren, J., 1969. Early results of the first large-scale steam soak project in the Tia Juana field, Western Venezuela. JPT 21, 101–110. January.
Dubrovai, K.K., Sheinman, A.B., Sorokin, MA, et al., 1936. Experience of thermal exploitation in Chusovskih Gorodkah. Oil Ind. (4), 23–32.
Green D.W., Willhite, G.P., 1998. Enhanced oil recovery. Texas, USA.
Ismailov, F.S., Mehtiyev, U.Sh., Gasymli, A.M., 2011. Experience of thermal recovery methods application on Azerbaijan oil fields, Baku.
Kuhn, C.S., Koch, R.L., 1953. In-situ combustion-newest methods of increasing oil recovery. Oil Gas J. 52 (14), pp. 92-96,113-114.
Latil, M., Bardon, C., Burger, J., Sourieau, P., 1980. Enhanced Oil Recovery. Technip, Paris.
Matheny, S.L., 1980. EOR methods help ultimate recovery. Oil Gas J. 79–124. March 31.
Prats, M.A., 1978. Current apraisal of thermal recovery. JPT 1129–1136. August.
Ramey, H.J. A current review of oil recovery by steam injection. In: Paper WPC-12247 Presented at the 7th World Petroleum Congress, Mexico City, Mexico, 2-9 April 1967.

Eson, R.L. Downhole steam generator - field tests. In: SPE-10745-MS Presented at the SPE California Regional Meeting, San Francisco, California, 24–26 March, 1982.

Romero-Zerón, L., 2012. Advances in enhanced oil recovery processes. In: Romero-Zerón, L. (Ed.) Introduction to Enhanced Oil Recovery (EOR) Processes and Bioremediation of Oil-Contaminated Sites. ISBN: 978-953-51-0629-6.

Ruzin, L.M., Ursegov, S.O., 2005. Elaboration of thermal methods of development of permian-carbon reservoir of Usinskoye oilfield. Oil Ind. (2), 82–84.

Sheinman, A.B., Dubrovai, K.K., Sorokin, MA, et al., 1935. Experiments on underground gasification of oil reservoirs under environment conditions. Oil Ind. (4), 48–61.

Stovall, S.L., 1934. Recovery of oil from depleted sands by means of dry steam. Oil Wkly. 17–24. Aug. 13.

Surguchev, M.L., 1985. Secondary and Tertiary Methods for Enhancement of Oil Recovery. Nedra Publishers.

Winkler, A.K. The Exploitation of Oil-Fields of Extremely High Oil-Viscosity by Wells, Under the Aapplication of Thermal Energy. 3rd World Petroleum Congress, The Hague, Netherlands, 28 May-6 June, 1951.

Ternary diagrams and miscible displacements

9

Abstract

Ternary diagrams allow to plot (e.g. represent, show) three variables with the total constant sum of the coordinates (the grid construct) and then to combine the grid with an overly of other information on top of it. An example grid is shown on Fig. 9.1.

Ternary diagrams allow to plot (e.g. represent, show) three variables with the total constant sum of the coordinates (the grid construct) and then to combine the grid with an overly of other information on top of it. An example grid is shown on Fig. 9.1.

Ternary diagrams have many names like ternary plots, Finetti diagrams, triangle plots and simplex plots. Three variables change from zero to the maximum (the maximum can be 1, it can be 100%, it can be anything else) along the triangle sides. The overlay can be a phase with a defined properties, for instance.

The important constrain is that, as was said above, the sum of the coordinates of every point of the diagram is constant. For the shown diagram

$$A + B + C = 1$$

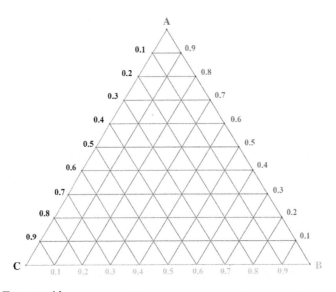

Figure 9.1 Ternary grid.

Primer on Enhanced Oil Recovery. DOI: https://doi.org/10.1016/B978-0-12-817632-0.00009-8

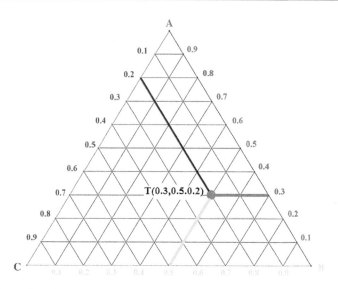

Figure 9.2 Drawing for the point T(0.3, 0.5, 0.2).

If, for instance, we will talk about composition then the sum of coordinates will be equal to 100%. The points on the diagram have coordinates **P(A, B, C)**. An example of point T(0.3, 0.5, 0.2) on the ternary grid is presented on Fig. 9.2.

While the grid is two dimensional, one or more additional layers are usually added to it. Those layers are used to reflect on the system of interest behavior. In a way, the diagram with layers becomes more complex and, at the same time, more useful.

Real life materials and processes are complex. In order to describe a processes and make the process model, we need to make some assumptions and simplification. In this method we ought to take onto an account only the most significant properties and processes in the system and then add less significant properties and processes only if our model do not predict behavior of the system accurately enough.

We have said before that the hydrocarbons in the reservoir (e.g. crude) are complex mixture of various hydrocarbon and other chemical compounds. From the view of the oil phase behavior, it is possible relatively accurately to describe crude properties if all hydrocarbons are divided into three groups according of number of carbons (heavy atoms) in the molecule:

1. Group one contains methane and is similar in properties for the oil extraction to carbon dioxide and nitrogen; it is named C_1 pseudo-component
2. Group two contains intermediate hydrocarbons with two to six carbons — C_2–C_6 pseudo-component
3. Group three contains heavy hydrocarbons — C^{7+} pseudo-component

As we have pseudo-components now the resulting diagram is named **pseudoternary diagram**. A hypothetical crude pseudoternary diagram is presented on Fig. 9.3.

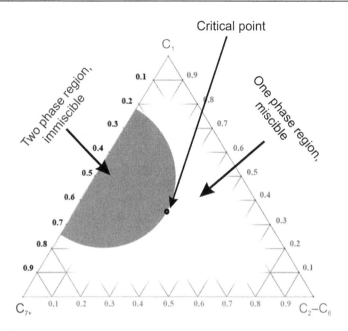

Figure 9.3 Pseudoternary crude diagram.

In order to provide process analysis an overlay of phase behavior is placed over the ternary grid. The overlays can be very different. The overlays might look very similar but can show different properties. One needs to pay attention to the information presented and discussed, as the diagrams might look simple and familiar but drawn for various purposes.

Few words should be said at this point. A pseudoternary diagram is drawn to provide process analysis and it is some model of the process, let us say mixing. As a model it needs to contain only significant parameters and the diagram is a simplification. The diagram usually shows at least two regions (sometimes there are more regions) in which liquids behave quite differently (see Fig. 9.3). Firstly, there is one phase, miscible, region. In this region the crude is uniform and can be relative accurately described as a single phase. Secondly, there is a two-phase (or multiphase) region. Inside the shaded area two (or more) immiscible phases co-exist. Above The system rheology need to be analyzed as at least two phase system. It is a tradition to mark on the introductory texts the critical point (also known as plait point). It is not worth going into the discussion on the system behavior at this critical point in this text at the level of introductory description. We will only say that in some conditions and systems this point may not exist.

Important fact is that a ternary diagram is drawn for the well-defined thermodynamic conditions. As thermodynamic conditions (temperature and/or pressure for instance) change the phase diagram changes as well. The diagram will be different if conditions have changed.

Generally, the higher is the pressure the smaller is two phase (or multiphase) region. This is demonstrated at the Fig. 9.4. It is also possible to say, but this is a less accurate statement compared to the pressure effect, that the temperature rise has an opposite effect for phase behavior − at higher temperature there is higher probability of multiphase region existence.

Pseudoternary hydrocarbon diagrams are important for analysis of many processes in oil reservoir behavior. The diagrams are used for oil, water, gas and surfactants behavior. The diagrams are either drawn on the results of laboratory investigations or by various modeling software.

Injected into reservoir fluids and gases somehow will interact with the crude. The exact interaction will define, at the end, the efficiency, in extraction terms, of the injection. It is possible to divide all interactions into two groups − immiscible and miscible. We will not talk about the emulsification at this point as the process needs different description model.

At the immiscible situation, two liquids/gases will not form a single phase. Some insignificant number of molecules will cross the border between two substances, nevertheless. "Insignificant" means that the number of those molecules would not be big enough to alter the phase behavior.

Miscible situation is more complicated. Pressure plays very important role. It is possible that the injected liquid (gas) and oil (reservoir hydrocarbons) will mix immediately on the first contact. This is so named First Contact Miscibility (FCM). It is close to impossible to achieve and high(er) pressure helps in this respect. LPG by many is regarded as First Contact Miscible liquid with a reservoir oil.

Most readily (for a broader spectra of substances) the situation is realised when the injected gas (liquid) mixes with the crude gradually through the process of exchanging components. This is so named Multiple Contact Miscibility (MCM). Some components from the injected gas (liquid) will dissolve in crude first, this will modify the oil and allow other components to dissolve. Some components of oil (light hydrocarbons) will dissolve in the injected gas (liquid). The gas (liquid) will be enriched with the oil light components. The process will continue until the single phase will be created.

Oil extraction is a dynamic situation. We need to make the reservoir liquids (we prefer oil) in the reservoir to move towards production wells so we would get the liquids to the surface. As liquids and gases in reservoir move the properties of substances at each physical point change dynamically.

Physical and chemical processes in many cases depend on concentrations and defined by the dominant process. It can be that oil light components predominantly evaporate into injected gas. In this case the injected gas in enriched with the oil light fractions. As the injected gas propagates through the reservoir, we can say that the mixing develops on the front zone and this drives virgin oil displacement. This process dominates oil displacement in the case of displacement by lean gas (methane and ethane) and it is named a vaporizing drive (see Fig. 9.5A).

If enriched gas injected to drive the displacement, then heavier hydrocarbons from the gas dissolve in oil and the mixing develops at the back of propagating gas slug. This process is named a condensing drive.

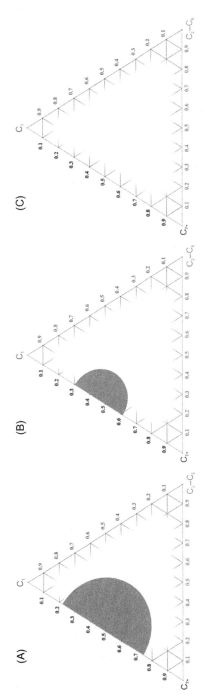

Figure 9.4 Effect of pressure increase on phase presence. Low pressure for example (A). Intermediate pressure for example (B). High pressure for example (C).

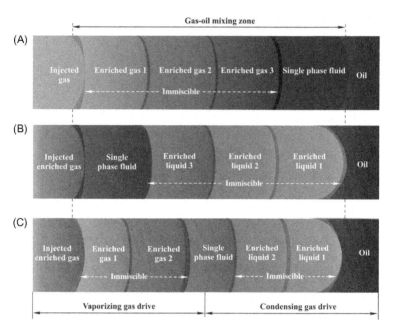

Figure 9.5 (A) Vaporizing drive; (B) − condensing drive; (C) − combined (real) drive.

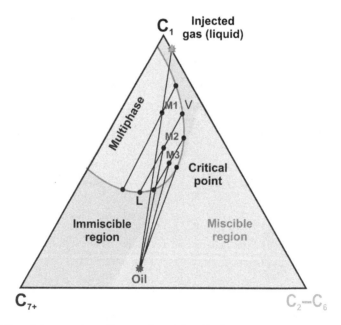

Figure 9.6 Pseudoternary phase diagram for mixing displacement.

The above model is a simplification. Majority of real displacement drives with carbon dioxide and enriched gas have both processes. Only processes during injection of dry hydrocarbons, nitrogen and flue gases are proceed by vaporizing scenario.

One of the areas where ternary diagrams are used extensively is the description (analysis) of reservoir oil behavior during various liquids and gasses injection into the reservoir. It is complicated area in oil extraction with many variables and processes. We will only introduce the topic here and then provide few more examples later in the book.

The vaporizing mixing can be analyzed as presented on Fig. 9.6. Injected gas (liquid) is going to mix with oil. At the initial mixing the mixture has the composition M1.This composition will separate on vapor (V) and liquid (L). The line connecting those states is named a tie or a connode. The vapor now contains more intermediate hydrocarbons. The vapor spreads through the reservoir faster than the liquid. Vapor again mixes with oil and created composition M2. The process with separation is repeated again, the vapor will get the mixture M3 and eventually will a to a critical point composition and will become a single phase miscible liquid.

We will discuss more specific examples in Chapter 10.

Gas flooding

10

Abstract

Undoubtedly, gas injection is one of the oldest methods in Enhanced Oil Recovery and its use constantly increases. Most of the field implementation of the method relates to the use of non-hydrocarbon gases — carbon dioxide, nitrogen and flue gases.

Chapter Outline

Undoubtedly, gas injection is one of the oldest methods in Enhanced Oil Recovery and its use constantly increases. Most of the field implementation of the method relates to the use of non-hydrocarbon gases — carbon dioxide, nitrogen and flue gases.

Mixing displacement by hydrocarbon gases can be divided into three separate methods: injection of liquefied gas or solvent; injection of enriched gas and injection of high pressure gas. Field tests and full-scale application have been carried out with all of them. In order to achieve miscibility with oil of liquefied gas or solvent, enriched gas and high-pressure gas, a sufficiently high pressure is required to get miscibility. If the reservoir conditions are not favorable for the gas flooding,

Primer on Enhanced Oil Recovery. DOI: https://doi.org/10.1016/B978-0-12-817632-0.00010-4

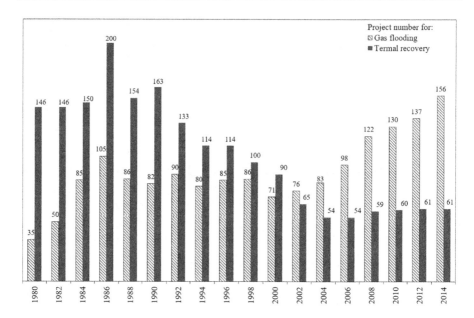

Figure 10.1 Project numbers of thermal recovery and gas flooding in USA.

then an early breakthrough of the working agent occurs and the sweep efficiency will be low. Hydrocarbons have been and are used widely for the gas injections too. However, hydrocarbons are valuable consumer products and there is an increasing reluctance to inject them back into the reservoir. In recent years, especially in the United States, the emphasis has shifted to less valuable, non-hydrocarbon gases, such as CO_2, nitrogen and flue gases. Although nitrogen and flue gases displace oil less efficiently than hydrocarbon gases, overall economic outcome can be somewhat more favorable.

A significant influence on the process of oil displacement by gas injection into the formation is exerted by its miscibility with oil. As a rule, at atmospheric pressure and room temperatures dry natural gas (methane) and oil do not mix. To ensure the mixing of oil and gas, the existence of a transition zone between them is necessary. In this case, the hydrocarbon mixture forming the transition zone must be dissolved in both gas and oil. This is usually achieved by enriching the injected gas.

The displacement of oil under conditions when complete mixing with the gas does not occur is called immiscible displacement. The best results in enhanced oil recovery by the gas flooding are observed when the oil is displaced by gas in a miscible displacement.

The CO_2 use for EOR significantly developed from the first patent for the method in 1952. Analysis of the EOR gas methods use of in the United States shows that there are periods during which interest in this method of EOR has increased significantly (see Figs. 10.1 and 10.2). The first period was during

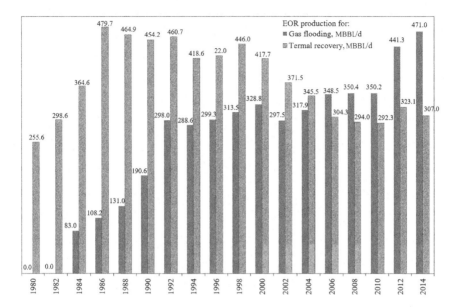

Figure 10.2 EOR Production of thermal recovery and gas flooding in USA.

1980–1986, when the number of implemented projects increased due to an increase in the oil price. During this time, the number of projects increased from 35 to 105, and the oil production increased to 108.2 thousand barrels per day.

The second period began in 2002, also due to a significant increase in the price of oil. During this period, the number of projects increased from 76 (2002) to 156 (2014), and production from 297.5 to 471 thousand barrels per day. At the same time, in 2006, for the first time, oil production by the implementation of gas methods exceeded the oil production obtained by the use of thermal methods (see Fig. 10.2).

10.1 CO$_2$ injection

The use of carbon dioxide flooding to increase oil recovery started in 1950 and has been very successful. This success is firmly based on many laboratory studies, field trials and the application experiences.

10.1.1 Background processes

In order to understand why carbon dioxide became an important working agent for injection into oil reservoirs, we must reflect first on the gas main properties and factors that make CO$_2$ a useful tool for the enhanced oil recovery.

Carbon dioxide is a colorless, odorless, inert and non-combustible gas. It has a molecular weight of 44.01. This is one and a half times higher than molecular

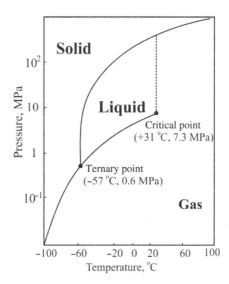

Figure 10.3 Carbon dioxide phase diagram. Temperature and pressure values are rounded.

weight of air. The phase diagram of carbon dioxide is shown in Fig. 10.3. Please note that the values were rounded

Carbon dioxide is highly soluble in oil and soluble, to a lesser extent, in water. At the same time, the following properties are known which enable carbon dioxide use in enhanced oil recovery when carbon dioxide mixes with the reservoir liquids:

- decrease in viscosity of crude oil and increase in viscosity of water;
- oil swelling and oil density reduction;
- multiple contact miscibility with hydrocarbons;
- acidic type interaction with the formation carbonates and clays.

As can be seen from Fig. 10.3, at a temperature below 31 °C and at high enough pressure, carbon dioxide exists in liquid phase. In the gas mixture with some hydrocarbons the critical temperature can rise up to 40 °C. Above the critical temperature and pressure the gas will behave as a supercritical phase.

For pure carbon dioxide the critical pressure is 7.3 MPa (approximately 170 atm or 2495 psi). At a pressure below the critical carbon dioxide would exist in a gaseous state. At a pressure of 8–25 MPa and a temperature of 20–1000 °C, the density of carbon dioxide in the liquid state is 500–900 kg/m^3, the viscosity is 0.05–0.1 mPas. The density of carbon dioxide at around 20 °C and atmospheric pressure is at around 2 kg/m^3. The viscosity is very low and it is in the region between 0.02 and 0.08 mPas (0.02–0.08 cP). This is more than ten times lower than the viscosity of water. If the pressure exceeds 15 MPa at a reservoir temperature below 40 °C, the density of carbon dioxide in the liquid and gaseous state are practically the same and are in the region 600–800 kg/m^3.

Dissolution in water. Carbon dioxide is much better soluble in water than hydrocarbon gases. At the same time, an increase in water salinity and temperature decreases the solubility of carbon dioxide. Pressure increase has the opposite effect and the solubility increases. Under reservoir conditions, the solubility of gaseous carbon dioxide in water is $30-60$ m^3/m^3 $(3-5\%)$.

The dissolution of carbon dioxide in water by $20-30\%$ increases water viscosity. When carbon dioxide is dissolved in water some carbonic acid is formed. The acid etches carbonates and clays. This etching opens and widens throats between formation grains and the permeability of carbonate rocks increases by $6-75\%$, and sandstone rocks by $5-15\%$. The acidic environment also reduces swelling of clays. This has a significant effect on increasing reservoir permeability.

Dissolution in oil. At the optimal conditions carbon dioxide has an excellent solubility in oil. Compared to water, oil can uptake $4-10$ times more of carbon dioxide at the optimal conditions. This high solubility also ensures CO_2 significant transfer of carbon dioxide to oil from an aqueous solution in oil-water contact. This transfer reduces the interfacial tension between oil and water, and the oil displacement becomes almost miscible. The highest mixing of carbon dioxide and oil occurs when the pressure of full mixing is exceeded, regardless of the CO_2 concentration. The specified pressure strongly depends on the physicochemical properties of the oil and is somewhere in the region $8-30$ MPa. At the same time, for heavy oils, high temperature and gas saturation pressure, the pressure of complete miscibility is significantly higher.

At pressures below the mixing pressure, carbon dioxide and oil separate, forming gaseous and liquid phases. In this case, the gas phase is formed by carbon dioxide with the light fractions of oil. The remaining liquid oil is stripped of light fractions. In this case it is possible that the liquid oil further separates into fractions and asphalt-resin-paraffin deposits (ARPD) start to precipitate and accumulate.

It should be noted that the viscosity of the oil is significantly reduced when carbon dioxide is dissolved in it. Separation of oil and carbon dioxide leads to significant increase in the reformed oil density and viscosity. This reformed oil is then left behind the front of propagating carbon dioxide slug.

Full mixing of carbon dioxide and oil at the beginning of carbon dioxide injection due to the above phenomena does not occur immediately. However, in the process of displacing oil, carbon dioxide is enriched with hydrocarbons, and the displacement becomes miscible. Therefore, the mixing pressure for carbon dioxide is substantially lower than for hydrocarbon gases, nitrogen and flue gases. For instance, the pressure required to displace light oil by mixing with hydrocarbon gases is almost two times higher than for carbon dioxide.

It should be noted that oil swelling (volume increase) with the dissolution of carbon dioxide has a significant effect on increase of oil recovery. When this occurs, a significant decrease in the viscosity of the oil is observed. The volume of oil increases $1.5-1.7$ times, while the increase in density is negligible $(2-3\%)$

10.1.2 Mechanism of the process

Mixing oil displacement. In the case of a miscible displacement, the oil is displaced by carbon dioxide like it is by a conventional solvent. In this case, three

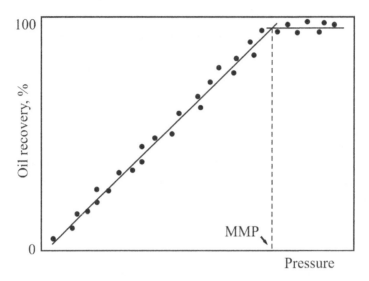

Figure 10.4 Minimal miscible pressure determination.

zones are formed in the reservoir in the direction of oil displacement: carbon dioxide zone; transition zone containing both CO_2 and oil; residual oil zone. Laboratory experiments on natural cores show that the displacement coefficient of the mixed displacement of oil by carbon dioxide can reach 0.95.

Measuring minimal miscibility pressure. The pressure of complete miscibility of carbon dioxide and the oil is determined experimentally. There are few methods to do it but the most common are two — gravitational-stable and slim tube tests.

In gravitational-stable test, a vertical formation rock filled column saturated with the oil first. The carbon dioxide is injected from the top. The gas displaces the oil vertically downward at a speed slow enough to maintain a gravitational-stable flow. The experiment is carried out at various pressures, and the coefficient of oil displacement is determined. A graph linking the oil displacement (recovery) with pressure is drawn (see. Fig. 10.4). As can be seen, the minimum miscibility pressure (MMP) is the pressure corresponding to the transition point to the maximum achievable recovery.

The second method for determining the minimum mixing pressure is an experiment in a slim tube. A spiral slim stainless steel tube with a diameter of 6 mm (0.25 inches) and a length of 1.2 m is packed with sand, saturated with oil at a given pressure and temperature. The sufficiently high value of the ratio of the length of the tube to its diameter provides a stable front for displacing oil with carbon dioxide without fingering. The results of the experiments are plotted and processed as above on the Fig. 10.4.

Immiscible displacement. In the case of immiscible displacement, the light fractions of oil are dissolved in carbon dioxide and part of the carbon dioxide is dissolved in oil. At the same time, due to the fact that carbon dioxide enriched in light fractions of oil displaces oil saturated with CO_2, component separation of oil happens, and difficult-to-recover oil saturated with heavy components is formed and left in the swiped zones.

At reservoir temperatures above the critical level, carbon dioxide is in a gaseous state regardless of the pressure. In this case, the development of the field will be much less effective than in mixing displacement. This is due to the unfavorable ratio of oil and gas mobility which leads to low sweep efficiency. In order to achieve technological and economic efficiency in the injection of carbon dioxide, it is necessary to ensure that it is in a liquid state. Only in this case the overall efficiency of the displacement is at its maximum. For this reason only reservoirs with temperatures close to critical $(25-35\ °C)$ will demonstrate the very best results.

Swelling effects. Carbon dioxide injection reduces mobility of water and increases the mobility of oil. This increases swipe efficiency by improving stability of the displacement front. In addition, oil volume swelling is one of the most important factors for oil displacement by carbon dioxide injection. The exact value of oil swelling (expansion) is a complex function of the light hydrocarbons content in the oil, reservoir pressure and temperature,

Bigger oil volume leads to artificial increase in oil saturation and to an increase in pore pressure. This leads to an effective displacement of residual oil. Just as the result of this process the oil recovery coefficient may increase by $6-10\%$.

10.1.3   Applicability criteria

Carbon dioxide injection should be carried out in reservoirs with moderately light oil (API ≥ 28), and the reservoir should be sufficiently deep (≥ 1500 m) to provide a sufficiently high pressure to achieve miscibility. Carbon dioxide, when dissolved in water, would reduce the interfacial tension between oil and water. However, produced acidity can also lead to problems with metal corrosion. In this process, about $20-50\%$ of carbon dioxide slug is displaced by water. Water is usually pumped jointly with CO_2 in the water alternating gas mode (WAG). This improves the mobility ratio for the displacing fluids and oil. The criteria for the applicability of the method are shown in Fig. 10.5.

Carbon dioxide injection is the fastest growing method of enhanced oil recovery in the United States. The implemented projects continue to demonstrate good additional oil production. Carbon dioxide injection is used both as a secondary and as a tertiary method for oil recovery. However, the largest CO_2 injection projects are EOR implementations on fields that have been in long-term development. As a rule, this is done in places where waterflooding has been applied for many years.

The CO_2 injection technique has its problems too. The method application challenges include:

- Early CO_2 breakthrough due to its low viscosity;
- Metal corrosion in production wells;
- The need to use materials withstanding CO_2 environment;
- The need to separate CO_2 from the extracted fluids;
- The need to further compress CO_2 for re-injection;
- High CO_2 demand per unit of oil produced.

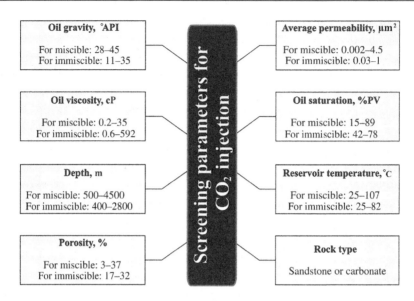

Figure 10.5 Screening parameters for CO_2 injection.

10.1.4 Implemented projects

Since the early 1970s, high oil prices have begun to generate widespread interest in carbon dioxide injection. In this case, both miscible and immiscible projects were used to displace oil by carbon dioxide.

Due to the presence in the United States of large reserves of carbon dioxide, which are located in the same geological basin with oil fields, this method of enhanced oil recovery is most prevalent in this country.

In order to cope with the world oil crisis in 1972, the US government passed a law on emergency energy security, which contributed to a significant increase in the US oil production. The research in promising EOR methods received strong development impulse.

After 1980, with an increase in demand for oil and gas, the price of oil and gas has significantly increased. The number projects for the injection of carbon dioxide experienced fast growth and in 2006 for the first time oil production obtained by this method exceeded oil production obtained by thermal methods of enhanced oil recovery. There were two big rises of interest in the method of carbon dioxide injection. The first rise occurred in the period 1980−1992, when laws and regulations were adopted that stimulated investment in the energy sector. The number of projects increased from 17 to 54, and the oil production increased to 144.97 thousand barrels per day (see Figs. 10.6 and 10.7). The second significant growth occurred in 2002, when the price of oil soared to $ 100 per barrel, which led to a significant increase in profits from the sale of CO_2 injection projects. During this period (after 2002) the number of projects increased from 67 to 137, and the oil production increased from 187.41 to 292.74 thousand barrels per day. As can be

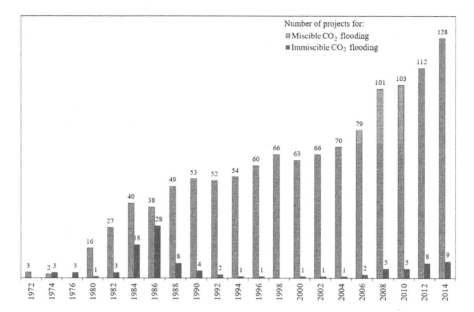

Figure 10.6 CO_2 EOR projects in the USA between 1972 and 2014.

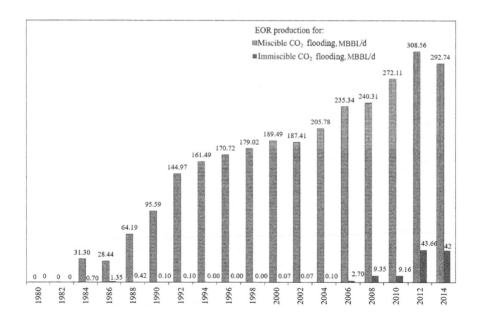

Figure 10.7 EOR output of CO_2 flooding.

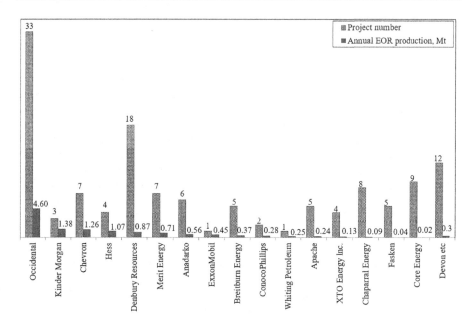

Figure 10.8 Miscible CO_2 flooding projects and annual production by a company in 2014.

seen from Fig. 10.6, in 2014 in the United States, 128 projects with mixed displacement were active, of which 104 were successful (81%). And only 9 projects were with immiscible displacement, which were less effective. At the same time, the main oil production (around 63%) falls on the 24 largest projects.

Fig. 10.8 enlists worldwide projects with the mixed CO_2 displacement for 2014.

Some valuable data can be extracted from the CO_2 injection projects data in the former USSR (see Fig. 10.9). It is useful to notice that almost in all cases increase in displacement efficiency is well correlated with the CO_2 slug volume.

The first field experiment in carbon dioxide injection in the USSR was carried out on Aleksandrovskaya Square of the Tuymazinsky deposit. The project started 1967 by the injection of carbonated water. Carbonated water was produced at the bottom of the well by mixing of simultaneously injected of carbon dioxide into the tubing and water into the annulus. As a result of the reservoir stimulation recovery increased by 15.6%.

10.1.5 Implementation technology

During the implementation of the carbon dioxide injection project, on the first place, it is necessary to calculate the projected demand in carbon dioxide. The produced volume then allows to assess necessary CO_2 site delivery methods and infrastructure.

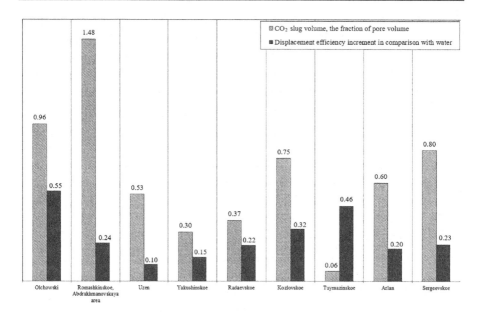

Figure 10.9 CO_2 flooding projects in the former USSR.

Volume of CO_2 for additional oil production is usually estimated during the planning stage. Roughly it is assumed that for every additional barrel of oil the demand for CO_2 will be:

- for secondary oil recovery of light oil at around 400 m^3 (0.8 t) and for heavy oil this rises approximately two times;
- for EOR of light oil it is around 500 m^3.

Later in this chapter we will provide an equation for the demand evaluation

A reliable CO_2 supply source is very important, since the gas must be supplied on a continuous basis in large volumes over long periods of time. It is expected that the project will last more than 5 years as the reservoir processes take some time to develop and the injection well spacing is usually significant. The gas used must have a purity of 90% or more. If impurities, other gases, are present, such as methane or nitrogen in high quantities a higher injection pressure will be required to provide miscible displacement.

The best sources are natural gas formations of high pressure with a high degree of purity. Usually such CO_2 natural formations are found in the search for oil and gas. In the United States, the oil producing basins of Wyoming, Utah, Colorado, and New Mexico have the largest reserves of carbon dioxide.

However, estimations show that currently known natural sources of high purity can provide only less than 15% of the total demand. Natural gas processing plants (for instance in Delaware and Val Verde basins in southwest Texas) produce significant amounts of carbon dioxide. In addition, all CO_2 exhausts from manufacturing can be used at the of energy "hungry" installations such as oil refineries, cement, ammonia, ethanol and ethylene plants. Power stations running on fossil fuel are another good sources. All this is more than enough to meet the existing demand.

The most convenient source should be considered for each specific project to ensure the economic efficiency of the injection project. Big gas quantities also demand an adequate form of transportation.

The method of transportation from the source to the oil field depends on the phase state of the carbon dioxide. Liquid requires smaller physical volumes but dictates right combination of temperature and pressure. At low injection rates (up to 0.2 million cubic meters per day), the least expensive method of transportation is to transport carbon dioxide as a liquid using existing insulated steel containers at a pressure of 2 MPa and a temperature of 0 °F (-17.8 °C). For large long-term projects, it is better to transport CO_2 by pipeline in a gaseous state (at a pressure of $10-14$ MPa which is somewhere above 100 atm). The conditions then ensure a single-phase flow.

The minimum mixing pressure needs to be experimentally determined for all reservoirs. Combination of all available data then allows to assess technological and economic efficiency of the method for a particular field and to begin the implementation.

There are several modifications of the enhanced oil recovery method by injection of carbon dioxide. Probably the less technologically demanding is an injection of carbonated water.

Injection of carbonated water is the easiest way to implement the method is to dissolve $3-5\%$ CO_2 in water and then inject the carbonated water into the formation (see Fig. 10.10). On the first place there is oil displacement by slightly more viscous water. At the same time good solubility of carbon dioxide in oil ensures its transfer from water solution to the oil remaining behind the displacement front. As a result, the interfacial tension between oil and carbon dioxide is significantly reduced, the volume of oil increases, the viscosity decreases, and the phase

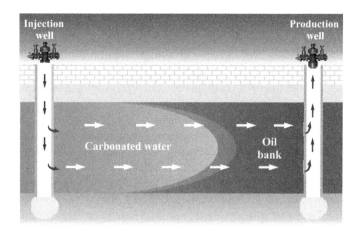

Figure 10.10 Injection of carbonated water.

permeability for oil increases accordingly. A significant lag in the front of carbon dioxide concentration from the displacement front lengthens the time for obtaining the CO_2 effect. Despite a slight increase in the sweep efficiency, as compared to the ordinary water flooding, the reduction of capillary forces and the improvement of the wettability, the final field recovery does not usually increase very significantly.

Injection of carbon dioxide slug is more effective and more technologically demanding method. The slug volume should be at around $10-30\%$ of the pore volume. The CO_2 slug is then followed by a water slug as shown on Fig. 10.11.

The processes in the reservoir during injection are complex, especially in mixed displacement volume. Somewhere in time after the carbon dioxide and water injections it is possible to try to artificially split the propagating displacement into number of zones and processes as it is shown on Fig. 10.12. The simplest zones are 1 and 6. In zone 1 oil in the virgin state is displaced by the injected liquids pressure towards the production well. Some connate water will be pumped out as well, but irreducible water saturation will remain. In zone 6 we have injected water and heavy oil which was stripped of light hydrocarbon fractions. Zone 2 contains moving

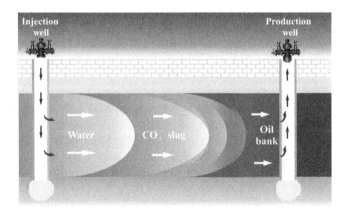

Figure 10.11 CO_2 slug injection method.

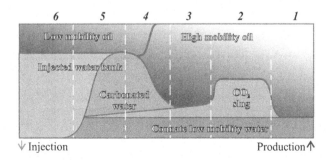

Figure 10.12 Artificial zoning and processes during CO_2/water slug injections.

carbon dioxide slug which actively mixes with oil. The oil, carbon dioxide and water are moving towards production well. In zone 3 carbon dioxide is fully mixed with oil, oil is swollen and has high mobility. This oil is pushed towards production well by advancing injected water in zone 4. Some carbon dioxide is transferred to the advancing water. In zone 4 we have movement of highly mobile oil and carbonated water. Somewhere in this zone all mobile oil is already displaced and only heavy, stripped of light hydrocarbons, oil is left behind. At the same time carbon dioxide is getting diluted by the advancing fresh water and at certain point zone 5 develops. Zone 5 has low mobility heavy oil, carbonated and modified towards fresh connate water.

There are different scenarios of the painted picture developing in time. What is clear that zones 1 and 2 will disappear and the rest of zones will spread in the formation volume with highly mobile liquids being displaced towards production well.

It has been shown that the optimal results are usually achieved when the CO_2 slug volume is at 20−30% of the pore volume. At the same time, the field production lifetime significantly increases, the volume of injected water, as compared to the simple water flooding, is reduced and high oil recovery factor is achieved. An important influence on the process has gravitational effect. In this regard, in the formations with high vertical permeability, the oil recovery factor would be significantly reduced (by a factor of 2−3).

Alternating injection of carbon dioxide and water is but a further method development (see Fig. 10.13) and according to the numerous studies has the highest efficiency.

In this modification effectiveness depends on the correct choice of the volume of the injected slugs and gas-water ratio. Injection of CO_2 and water can be done sequentially or simultaneously.

The implemented projects show that the optimal ratio of slugs of carbon dioxide and water lies in the range between 0.25 and 1. Low CO_2 volumes reduce the method efficiency and the effect of the implementation it closer to the efficiency

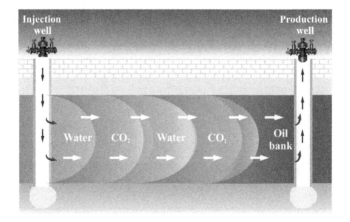

Figure 10.13 Alternating CO_2 − water injection.

when just carbonated water is injected. At the same time, too high volumes of CO_2 can lead to an early breakthrough of carbon dioxide into producing wells. In a heterogeneous reservoir, the volumes of the slugs should be lower than in a homogeneous reservoir. In the case of complete miscibility of carbon dioxide and oil, the volumes of the slugs can be quite large and reach 20% of the pore volume. At low miscibility, the volumes of the slugs should be rather small with the frequent alternation.

Other method modifications are developed in order to alleviate demand for high volumes of CO_2. This development stimulated search for the further improvement of the injection technique. To reduce the consumption of carbon dioxide, it is possible to use alternating injection of carbon dioxide, water and other more readily available gases. Also, it is possible to add various surfactants to water, which will ensure the formation of foam in the reservoir. It is also possible to use polymer additives to the injected water.

The field development process for the application of carbon dioxide injection practically does not differ from the system of development in the case of ordinary water flooding. Well placement is not significantly different from the conventional waterflooding. However, in a presence of high vertical permeability and the high gravity segregation of carbon dioxide and water, the well network should be densified.

Choosing the most economical method for the field development should include clear goal of the development and the planning ought to take into the account various factors of technical and capital investment. Significant volume of information needs to be gathered, including well documented history of the field. Planning for the flexibility in all aspects of injection and production is highly desirable. Eventual handling of big volumes of water and gas in the production wells needs to be prepared for. Re-injection needs to be undertaken as environmental concerns become more noticeable and CO_2 recycling is economically efficient.

It also needs to be taken into the account that the positive outcomes of CO_2 injection while certain would need time to develop.

10.2   Hydrocarbon injection

It was mentioned before that with the injection of hydrocarbon gases into the reservoir, oil can be extracted without mixing or with mixing. Dry natural gas at normal pressures and temperatures, even in deep oil strata, does not mix with oil easily. To ensure mixing, a transition zone is needed in which both oil and gas intermix. However, during the injection of dry hydrocarbon gases, the transition zone is not easily formed, and the oil is extracted without mixing. To ensure more efficient mixing, three different methods are used: high pressure gas injection; enriched gas injection and liquefied gas injection.

The mechanism of mixing displacement of oil by hydrocarbon gas:

- Creating miscibility when pumping high-pressure gas and enriched gas (this ensures zero interfacial tension between the oil and the working agent);

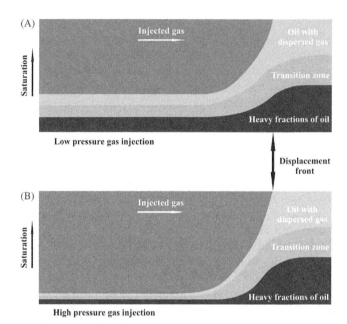

Figure 10.14 Oil displacement schemes for immiscible hydrocarbon gas injection.

• Increased oil volume (swelling);
• Reduced oil viscosity.

10.2.1 Process mechanisms

Processes during hydrocarbon injection depend on the formation conditions and oil composition. It is well known that full mixing is only achieved through formation or intermediate zone where both — injected hydrocarbons and oil undertake components exchange until the difference between the two complete disappear. In the absence of mixing zone the oil display network happens without mixing.

The processes without mixing are shown on Fig. 10.14. In order to emphasize the processes the oil is split into three components: light oil with dissolved gas, intermediate components and heavy fraction.

Low pressure injection has very limited gas dissolution in oil. It is possible to see that beyond the displacement front oil saturation is only reduced to a certain, still quite high level. The composition of remaining oil is not very different from the original oil in place.

High pressure injection leads to much lower remaining oil saturation after the displacement front as compared to low pressure injection. This low saturation is provided by few essential processes:

• the injected at high pressure gas has higher viscosity than the gas at low pressure;
• there is partial dissolution of injected gas which leads to reduction of oil viscosity;
• the oil is slightly swollen by the dissolved gas;

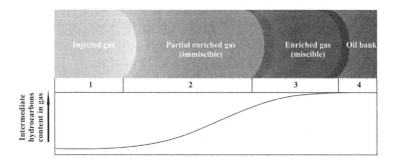

Figure 10.15 Oil displacement scheme for miscible hydrocarbon gas injection.

- the injected gas itself during propagation through the reservoir is enriched by oil components;
- the remaining oil slightly shrinks due to loss of the above components.

The effectiveness of the displacement process is significantly affected by the pressure. The bigger the injected gas dissolution, the bigger is displacement. This is mostly due to the oil swelling by the injected gas.

Displacement of oil with full mixing is demonstrated on Fig. 10.15. The injected gas should have pressure higher than the pressure of full mixing. In this case it is possible to distinguish four zones. The first zone contains injected gas. Partially enriched by oil components gas is forming the second zone. After significant component exchange oil and injected gas are undisguisable (Zone 3). The displaced virgin oil forms the last, forth, zone.

İt is also possible to see how the concentration of intermediate carbons increases from zone to zone. The concentration reaches the concentration of those hydrocarbons in the virgin oil on the border between third and fourth zones.

It is convenient to analyze the displacement process in pseudo-ternary diagram when the process of oil displacement. The analysis can be done as described in the previous chapter.

Fig. 10.16 shows a ternary phase diagram with boundary phase curves for three pressures. As can be seen from the figure, as the pressure decreases, the mixing zone decreases, and the two-phase region, on the contrary, increases. Considering the process of displacing oil of composition O by gas with composition G, we find that at 21 and 24.5 MPa, the miscibility of oil and gas is not achieved. With this composition of oil and gas miscibility occurs only at a pressure of 28 MPa. Studies have shown that to achieve mixing at a pressure of 28 MPa it is necessary that the concentration of intermediate components in the oil be at least 35%, otherwise higher pressure will be required.

Thus, the most important factor ensuring the miscibility is the composition of the oil, namely the presence in it of a sufficient amount of intermediate hydrocarbons. Only presence of intermediate hydrocarbons and high enough pressure guaranty, complete gas mixing with oil is achieved (i.e., all mixtures will be single phases).

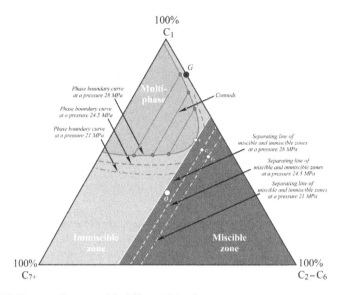

Figure 10.16 Ternary diagram with different injection pressure.

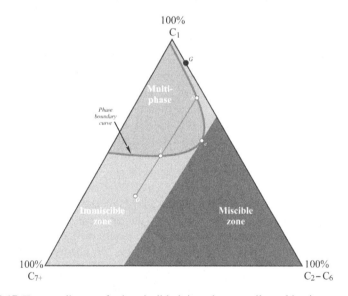

Figure 10.17 Ternary diagram for immiscible injected gas − oil combination.

Let us now consider the process of displacement under immiscibility conditions (see Fig. 10.17). As can be seen from the figure in the process of displacing oil by gas, the composition of the gas, although enriched, does not reach the composition of the critical (single phase) composition. In this case, the increase in the volume of oil does not occur, since the reservoir pressure does not exceed the saturation pressure and the gas does not dissolve in the oil. In this regard, an increase in oil recovery occurs due to an increase in the viscosity of the gas as a result of enrichment with intermediate hydrocarbons and a resultant decrease in the ratio of mobility of oil and gas.

Enriched with intermediate hydrocarbons (C3−C6), primarily propane, is often injected into the formation. Moreover, due to the fact that the concentration of intermediate hydrocarbons in a gas is higher than in oil, they begin to dissolve in oil, which leads to oil swelling and a decrease in viscosity. As a result, oil recovery significantly increases.

Liquefied gas injection started in the early 1950s. Numerous field tests and several full-scale implementations using liquefied petroleum gas were carried out in the USA for mixing displacement. However, the need for excessively large portions of liquefied petroleum gas and the high market value of liquid propane made the injection of liquefied petroleum gas economically unattractive. In this regard, to reduce the cost of the technology has been later modified.

A small slug of liquefied petroleum gas is injected into the formation, which forms a transition zone between the gas slug and oil. Usually, propane or its mixtures with other intermediate hydrocarbons are used as a slug of liquefied gas. The displacement becomes miscible. The main advantage of this method is the absence of capillary forces that prevent enhanced oil recovery when using immiscible methods. The efficiency of oil displacement this way increases significantly.

Injection pressure should ensure the maintenance of liquefied gas in a liquid state, complete miscibility of a liquefied gas, both with oil and with a dry propulsion gas. At the same time, the required mixing pressure is significantly lower than when injecting just dry or enriched gas. The need to maintain the liquefied gas in the liquid state limits the application of the method to deposits with a temperature below the critical for the gas used. For instance, the critical temperature for pure propane is 96.6 ^{0}C, and liquefied propane cannot be used at higher temperatures.

Fig. 10.18 shows a diagram illustrating the mechanism of oil displacement during the injection of liquefied propane. As can be seen from the figure, five successively moving zones are formed in the process of displacement: the first zone (from the injection well) is occupied by the injected dry gas; in the second zone the gas mixture moves with propane; the third zone is occupied by pure propane; a mixture of propane and oil moves in the fourth zone; the fifth zone is occupied by the displaced oil. This distribution of zones is due to convection and diffusion mixing. The liquefied propane moves behind the displacement front, in front of it and behind it there are transition zones containing its mixture with dry gas and oil.

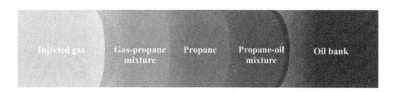

Figure 10.18 Oil displacement scheme for the liquid propane supported flooding.

As propane progresses through the reservoir, its concentration in the slug decreases. Over time, a significant decrease in the concentration of propane in the slug can lead to a significant decrease in the efficiency of the process. To ensure the efficiency of the displacement process, it is necessary to calculate the optimal size of the propane slug.

10.2.2 Applicability criteria

Injection of hydrocarbon gases should be carried out at sufficiently deep-lying formations. The minimum depth is determined by the pressure required to maintain miscibility. The required pressure ranges from about 8 MPa when pumping liquefied gas to 20−35 MPa when pumping high pressure gas, depending on the composition of the oil.

For the implementation of the process, steeply dipping seams are preferred, providing gravitational stabilization of the displacement front, which usually has an unfavorable mobility coefficient. The criteria for the applicability of the method are shown in Fig. 10.19.

The disadvantages of the method include:

- Viscous instability leads to low vertical and horizontal formation sweep efficiency;
- A large quantity of expensive consumable products is required;
- There is a significant loss of expensive injected material, which cannot be completely recovered later.

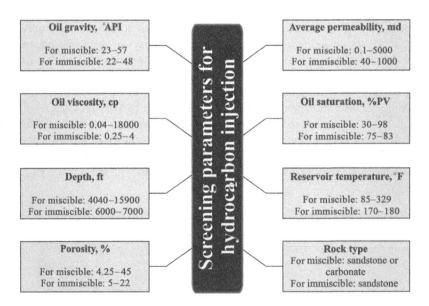

Figure 10.19 Screening parameters for hydrocarbon injection.

10.2.3 Implemented projects

The use of gas methods in the United States has two significant increases of interest in methods of enhanced oil recovery, including the injection of hydrocarbon gases (see Fig. 10.20). The first jump occurred in 1980, due to rising oil prices and continued until 1992. During this period, the number of hydrocarbon gas injection projects increased from 9 (1980) to 25 (1992). The oil production has risen to 113 thousand barrels per day. The second acceleration in the method implementation occurred after 2002, when oil prices soared and exceeded $ 100 per barrel. The high price generated a significant profit from gas methods of enhanced oil recovery. During this period, the number of hydrocarbon gas injection projects increased from 7 (2002) to 14 (2014), and oil production increased from 95.3 to 127.5 thousand barrels per day. It should be noted that this is significantly lower than oil production from the use of carbon dioxide injections, which on the same date (2014) amounted to 335.53 thousand barrels per day.

Fig. 10.21 shows the implementation of the method of hydrocarbon gases injection around the world. As can be seen from the figure, in addition to the United States, the specified method since 1969 has been implemented in Canada, Norway, Venezuela and the United Kingdom.

10.2.4 Implementation technology

During the implementation of the hydrocarbon injection project, first it is necessary to calculate the gas demand (dry or enriched) for the production of a unit of an

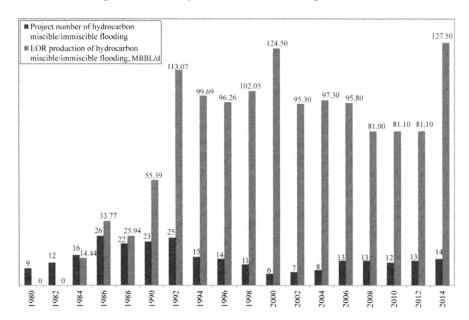

Figure 10.20 Hydrocarbon flooding EOR projects operating in USA between 1980 and 2014.

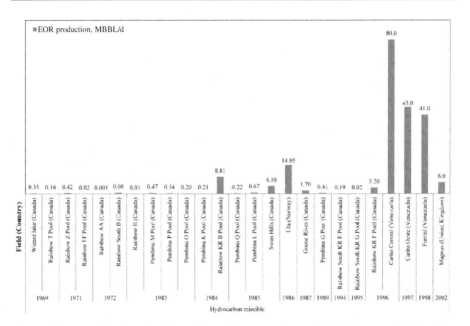

Figure 10.21 Hydrocarbon flooding EOR projects operating worldwide between 1969 and 2014.

additional oil. In this case, the calculation of the volume of injected gas is carried out with consideration for the reservoir pressure and temperature.

The injected gas rate is determined as follows:

$$V_{ig} = V_o + V_w + V_g,$$

where: V_{ig} is volume rate injected gas at the reservoir conditions, m^3/day; V_o is volumetric flow rate of produced oil at reservoir conditions, m^3/day; V_w is volumetric flow rate of produced water at reservoir conditions, m^3/day; V_g is volumetric flow rate of free produced gas at reservoir conditions ($V_g > 0$ only in the case of production wells operation with an accidental gas breakthrough from the gas cap into the production well), m^3/day.

The parameters included in the above equations are determined by the formulas:

$$V_o = V_{od}b_o$$
$$V_w = V_{wd}b_w,$$

where, V_{od} and V_{wd} are volumes of oil and water after degassing (under standard conditions), m^3/ day; b_o, b_w are volumetric coefficients of oil and water at reservoir conditions.

It can be that there is a enough of hydrocarbon gas on the site itself. In this case, gas need to be prepared locally for the injection. The gas is cleaned from

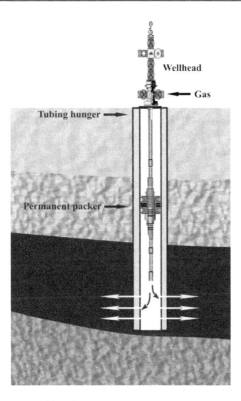

Figure 10.22 Gas injection well design.

mechanical impurities and compressed. Gas cleaning can be done with oil or cyclone dust collectors, as well as gravity separators. To compress gas, it is necessary to install a compressor station, the capacity of which is calculated from the required volume of hydrocarbon gas and injection pressure. If high-pressure gas is present in the field, injection can be carried out in an uncompressed manner, which significantly reduces capital and operating costs.

The injection of hydrocarbon gases is carried out through a separate tubing, which is usually lowered to the upper holes of the well filter. The annular space is blocked by a drilled packer, which is lowered onto the well. This protects the casing from high gas injection pressure. The design of the injection well is shown on Fig. 10.22.

A serious problem with the technology implementation is events of gas break-throughs in the production wells. To prevent a breakthrough, it is necessary to control the gas injection rate, the gas factor and the composition of the produced gas. In case of a breakthrough it is essential to stop it. For this it is possible to lower fluid withdrawal from the wells.

Modifications to the method can be applied to reduce the consumption of valuable hydrocarbons. For instance, after injecting the slug of enriched gas in a volume enough to effectively displace the oil, it is possible to inject cheaper gases. For instance, it is possible to use nitrogen and flue gas. Flue gas, as is known, consists of almost 90% of nitrogen and about 10% of carbonic acid. On the negative side, the carbonic acid will dissolve in connate water and the displacement front will be only then provided by nitrogen. However, if the injection pressure is high enough, and the oil contains a sufficient amount of intermediate hydrocarbons, then the nature of the displacing agent is not so important.

In cases, when it is possible to inject flue gases after hydrocarbon gases, the volume of the hydrocarbon slug can be as low as 5% of the pore volume. An important advantage of flue gases is a large compressibility factor, which significantly reduces the financial cost of compressing gas. The disadvantage of flue gases is that they must be supplied to the field in a compressed form.

When implementing the injection of enriched hydrocarbon gas, it should be borne in mind that it is rarely when a field produces a sufficiently fat gas and it is necessary to obtain enriched gas by mixing dry natural gas with liquefied gas. In order to reduce the consumption of such a valuable commodity it is necessary to find the permissible dilution of liquefied gas with a dry hydrocarbon gas at the discharge pressure and reservoir temperature before starting the process.

The system of field development in the injection of hydrocarbon gases is practically the same as the system of development in the case of the conventional waterflooding. The main requirement for the system development is that the injection should be produced in an areal configuration. Well placement is also not significantly different from waterflooding. However, with high vertical permeability and the presence of high gravity segregation of hydrocarbon gas and water, the well network should have higher density.

10.3 Nitrogen and flue gases injection

To replace expensive hydrocarbon gases, relatively inexpensive nitrogen and flue gases can be used as an oil displacer. Depending on the injection pressure and composition of the oil, the displacement process can be miscible and immiscible. Due to their low cost, large volumes of these gases can be introduced into the formation. Nitrogen or flue gases can also be used to move the hydrocarbon gases or carbon dioxide slug. All this significantly reduces the cost of the project.

The mechanism of oil displacement with nitrogen and flue gases:

- at a sufficiently high injection pressure, evaporation of light fractions of oil occurs, which provides a miscible displacement;
- ensuring gas-pressure regime, in which a significant part of the reservoir volume is filled with relatively inexpensive non-hydrocarbon gas.

The advantages of nitrogen and flue gases include low cost and availability, low compressibility (3 times less compared to carbon dioxide and 1.5 times less compared to methane), which provides significantly lower costs for compression (2−3 times less). Nitrogen has further advantage as it does not corrode, as opposed to flue gases, the metalworks.

The disadvantages include the low solubility of nitrogen in oil ($35-45$ m^3/m^3 for light oil and $15-25$ m^3/m^3 for heavy oil). The solubility of flue gases is not so different from nitrogen as the gas consists by almost 90% percent of nitrogen. The minimum mixing pressure for nitrogen exceeds 35 MPa, which is significantly higher than for natural gas (25 MPa) and carbon dioxide (8 MPa).

10.3.1 Process mechanisms

Depending on the oil type, reservoir conditions and the injected gas type (composition) an oil/gas front propagation and interaction can proceed in evaporation mode or in condensing gas mode.

10.3.2 Applicability criteria

Miscible displacement during injection of nitrogen and flue gases only should be carried out in deep oil deposits. The reason was outlined just before − mixing is only achieved at high pressures.

The displacement front has unfavorable mobility ratio. For this reason, as with the injection of hydrocarbon gases, the provision of gravitational stabilization of the displacement front is only possible in steeply falling oil strata. The criteria for the applicability of the method are shown in Fig. 10.23.

The disadvantages of the method include:

* viscous instability leads to low vertical and horizontal formation sweep efficiencies;
* corrosion can cause problems during flue gas injection;

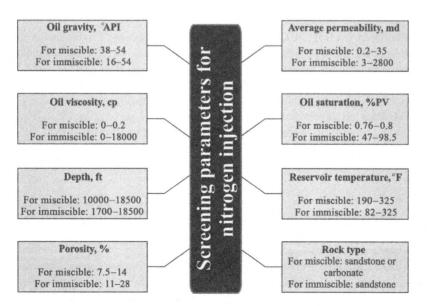

Figure 10.23 Screening parameters for nitrogen and flue gases injection.

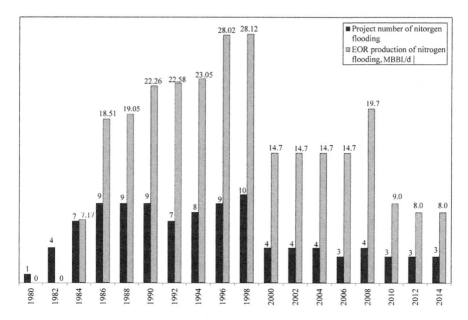

Figure 10.24 Nitrogen flooding EOR projects operating in the USA between 1980 and 2014.

- increased costs on-site natural gas purification as non-hydrocarbon gases must be separated at the end from the hydrocarbons.

10.3.3 Implemented projects

It is possible to see (Figs. 10.24 and 10.25) that the numbers of implemented in the US projects for the injection of nitrogen and flue gases are relatively small. The maximum number for nitrogen injection, just 10, was in 1998. At the same time, additional oil production amounted to 28.12 thousand barrels per day. After 1998, the number of projects implemented in the USA has been significantly decreasing and additional oil production dropped to 8000 barrels per day in 2014. For flue gases, only two projects were implemented in 1992 and production was 11,000 barrels per day. The amount of additional oil production obtained from the injection of nitrogen and flue gases is significantly lower than for carbon dioxide and hydrocarbon gases.

10.3.4 Implementation technology

When implementing nitrogen and flue gases injection project, as well as for other gases, first of all, it is necessary to calculate the gases volume for the extraction of a unit of additional oil. In this case, the calculation of the volume of injected gas is carried out by taking in to the account the reservoir pressure and temperature. The calculation can be carried out according to the formula shown in Section 10.2.4.

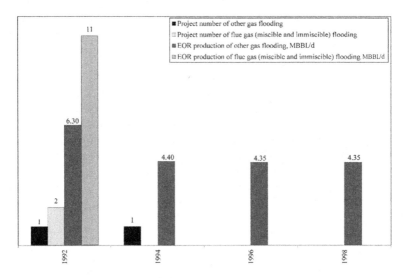

Figure 10.25 Flue and other gases flooding EOR projects operating in USA between 1992 and 2014.

According to the accumulated expertize, for extraction of one barrel of extra light oil, about $45-70$ m^3 of nitrogen are needed.

Nitrogen production is carried out from the air by cryogenic rectification (cryogenic air separation) or using non-cryogenic membrane air separation technology. Cryogenic technology is most common for nitrogen production at the field. The capacity of the plant, depending on the need, is at around $0.3-1.5$ million m^3/day. Potential capacity of a larger installation can be as high as 6 million m^3/day. The largest cryogenic plant for producing nitrogen from air was built in 2000 and enlarged in 2004 to pump nitrogen into the reservoir in the Mexican Cantarell and Ku Maloob Zaap fields located in the Gulf of Mexico. The capacity of this installation is about 43 million m^3/day (50,000 tons/day). Typically, a cryogenic plant contains air compressor to supply air, the cryogenic distillation plant itself and compressors to deliver nitrogen for the injection.

The injection well construction is like the shown earlier for the other gases.

There are few nitrogen applications types.

Immiscible displacement. Injection into to gas cap. Nitrogen is injected into reservoirs that have significant gas caps (see Fig. 10.26). Nitrogen is injected in this case to maintain reservoir pressure and extract hydrocarbons. Nitrogen get gravity separated from the gas cap and oil and improves extraction of both oil and hydrocarbons gas

Reservoir pressure maintenance. Nitrogen is can be introduced into the gas condensate formation under high pressure to maintain reservoir pressure (see. Fig. 10.27).

Nitrogen injection allows to keep high pressure, maintain and to restore phase permeability. All this provides high gas condensate recovery. The process is carried

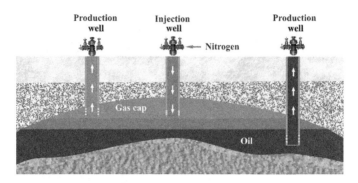

Figure 10.26 Gas cap displacement.

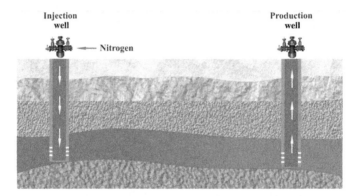

Figure 10.27 Gas condensate pressure maintenance.

out at fields with a reservoir pressure close to the pressure of the onset of condensation, high rate of liquid precipitation during condensation (20–40% of the pore volume) and with a significant condensate yield of at least 10^{-3} m^3/m^3.

Miscible displacement. The miscible displacement of oil during the injection of high-pressure nitrogen is provided by the evaporation of intermediate hydrocarbons from oil and their dissolution in nitrogen. This forms a transition zone mixing, containing a mixture of nitrogen and intermediate hydrocarbons, which is soluble in oil. This process effectively displaces oil to producing wells (Fig. 10.28).

In some cases, after the injection of nitrogen a slug of water is injected. Eventually, this forms a slug with nitrogen/water mixture. It has been shown to provide an efficient oil displacement of oil (Fig. 10.29).

Gravity drainage is used in reservoirs with the good potential for the nitrogen injection process due to the large depth or strata inclination. Nitrogen, which has a lower density than the reservoir oil, when it is introduced into the top of the reservoir, will displace the oil under it (see Fig. 10.30). In this case the displacement is piston-like without any displacement front viscosity instabilities. In this case the greater the slope the higher is interface stability. The gas distribution is

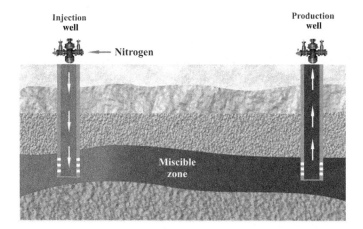

Figure 10.28 Miscible displacement by nitrogen injection.

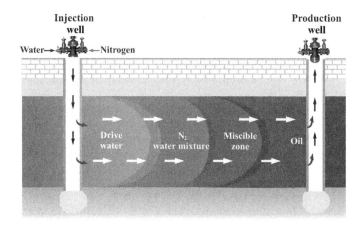

Figure 10.29 Nitrogen/water miscible displacement.

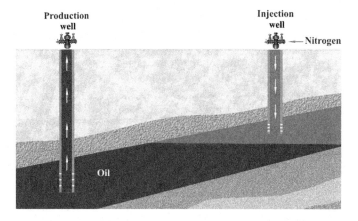

Figure 10.30 Gravity drainage with nitrogen injection.

predominantly guided by the gravity. The injection rate is high and slope dependent. At high inclination the injection rate can be very high.

The system of field development for nitrogen injection is the same as for water flooding with the suggestions about injection well grid densification if there is high vertical permeability.

Further reading

Alagorni, A.H., Yaacob, Z.B., Nour, A.H., 2015. An overview of oil production stages: enhanced oil recovery techniques and nitrogen injection. Int. J. Environ. Sci. Dev. 6 (9), 693−701.

Alvarado, V., Manrique, E., 2010. Enhanced Oil Recovery: Field Planning and Development Strategies. Gulf Professional Publishing.

Anada, H., Sears, J., et al. Feasibility and Economics of By-Product CO2 Supply for Enhanced Oil Recovery. Final Report, vol. 1. Technical Report. DOE Contract No. DE-AT21-78MC08333-3, U.S. Department of Energy, Bartlesville, OK, January 1982, pp. 96−98.

Balint, V. The Use of Carbon Dioxide in the Oil Production. Moscow, Nedra, 1977, 240 p.

Bath, P.G.H., van der Burgh, and Ypma, J.G.J.: Enhanced oil recovery in the North Sea, In: Proc. 11th World Pet. Cong. London, ATD 2 (2), (1983).

Benham, A.L., Dowden, W.E., Kunxman, W.I., 1960. Miscible fluid displacement prediction of miscibility. J. Pet. Technol. 12 (10).

Booth, R., 2008. Miscible Flow Through Porous Media. University of Oxford.

Brock, W.R., Bryan, L.A. Summary results of CO2 EOR field tests, 1972−1987, SPE 18977. In: Paper Presented at the SPE Joint Rocky Mountain Regional/Low Permeability Reservoirs Symposium and Exhibition, Denver, Colorado, March 6−8, 1989.

Carcoana, A., 1992. Applied Enhanced Oil Recovery. Prentice Hall, Inc, New Jersey, USA.

Clancy, J., et al., 1985. Analysis of nitrogen-injection projects to develop screening guides and offshore design criteria. J. Pet. Technol. 37 (6), 1097−1104.

Crump, J.S. Method of Increasing Recovery of Oil. Brit. P. No. 868649, 25 May 1961.

Daltaban, S.T., Lozada, A.M., Pina, A.V., Torres, F.M. Managing water and gas production problems in Cantarell; a giant carbonate reservoir in Gulf of Mexico, SPE-117233-MS. In: Paper Presented at the Abu Dhabi International Petroleum Exhibition and Conference, Abu Dhabi, UAE, 3−6 November 2008.

Farias, M.J., Watson, R.W., 2007. Interaction of Nitrogen/CO2 Mixtures With Crude Oil. Pennsylvania State University, University Park, PA, USA.

Green, D.W., Willhite, G.P. Enhanced Oil Recovery, Richardson, Tex.: Henry L. Doherty Memorial Fund of AIME, Society of Petroleum Engineers, 1998.

Holm, L.W., 1976. Status of CO2 and hydrocarbon miscible oil recovery methods. In: J. Pet. Technol.

Holtz, M. Immiscible Gas Displacement Recovery, PRAXAIR, 2012.

Jishun, Q.I.N., Haishui, H.A.N., Xiaolei, L.I.U., 2015. Application and Enlightenment of Carbon Dioxide Flooding in the United States of America. Pet. Explor. Dev. 42 (2), 232−240.

Johns, R.T., Dindoruk, B., Orr Jr., F.M., 1993. Analytical solutions for dispersion-free flow of two-phase, four-component mixtures confirm the existence of condensing/vaporizing gas drives and reveal how they behave. SPE-24112. SPE Adv. Technol. Ser. I (2).

Juttner, I., 1997. Oil displacement in miscible condition. Rudarsko-Geološko-NAFTNI Zbornik 9 (1), 63−66.

Koch, H.A., Hutchinson, C.A., 1958. Miscible displacements of reservoir oil using flue gas. J. Pet. Technol. 10 (1).

Koch, H.A., Slobod, R.L., 1957. Miscible slug process. Trans. AIME. 210, 40.

Kokal, S., Al-Kaabi, A., 2010. Enhanced Oil Recovery: Challenges & Opportunities. World Petroleum Council: Official Publication.

Koottungal, L., 2008. Worldwide EOR survey. Oil Gas J. 47−59. Apr. 21, 2008.

Koottungal, L., 2014. Worldwide EOR survey. Oil Gas J. 79−91. Apr. 7, 2014.

Lake, L.W., 1989. Enhanced Oil Recovery. Prentice Hall.

ISBN: 978-0-7506−7785-1 second ed. Lyons, W., Plisga, B.S. (Eds.), 2005. Standard Handbook of Petroleum & Natural Gas Engineering, 13. Elsevier Inc, Burlington, MA.

Moritis, G., More, U.S., 2008. EOR projects start but EOR production continues decline. Oil Gas J. 41−46. Apr. 21.

Pautz, J.F. et al. NIPER-471 Review of EOR Projects Trends and Thermal EOR Technology − Topical Report. Performed under cooperative agreement No. FC22-83FE60149-DOE 1990 IIT Research Institute, NIPER/DOE Bartlesville, OK

C.M. Quintella, C. de Ondina, R. Dino, A.P. Santana Musse. CO2 Enhanced Oil Recovery and Geologic Storage: an Overview with Technology Assessment Based on Patents and Articles. SPE 126122. Paper presented at the SPE International Conference on Health, Safety and Environment in Oil and Gas Exploration and Production held in Rio de Janeiro, Brazil, 12−14 April 2010.

Physico-chemical studies of systems and materials based on rare elements. In: Serebryakov, Y.A. (Ed.), Collection of Scientific Works. KNC SA USSR, Apatity, 96 p.

Sheng, J.J., 2013. Enhanced Oil Recovery Field Case Studies. Elsevier Inc, UK, p. 683p.

Shine, J., Holtz, M. Reserve growth & higher recovery using nitrogen gas injection. In: Proc. 2008 Wyoming EOR/IOR Conference, The Wyoming Enhanced Oil Recovery Institute's: PRAXAIR. Inc, 2008.

Stalkup Jr., F.I., 1984. Miscible Displacement. SPE Monograph Series. SPE, Richardson, TX, pp. 137−156.

Surguchev, M.L., 1985. Secondary and Tertiary EOR Methods. M.: Nedra.

Surguchev, M.L., Gorbunov, A.T., Zabrodin, D.P., 1991. Residual Oil Recovery Methods. M.: Nedra.

Taber, J.J. Technical screening guides for the enhanced recovery of oil. In: Paper SPE-12069-MS. SPE Annual Technical Conference and Exhibition, 5−8 October 1983, San Francisco, California.

Taber, J.J., Martin, F.D., Seright, R.S., 1997. EOR screening criteria revisited − Part 1: Introduction to screening criteria and enhanced recovery field projects − Part 2: Applications and impact of oil prices. SPE Reservoir Eng., pp.189−198; 199−205.

Terry, R.E., 2000. Enhanced Oil Recovery, pp. 503−518.

Van Dyke, K., 1997. Fundamentals of Petroleum, fourth ed. The University of Texas at Austin, Austin, Texas.

Van Poollen, H.K., Associates, 1980. Fundamentals of Enhanced Oil Recovery. PennWell, Tulsa, OK, p. 133.

Walker, D., 2008. Reserve Growth & Higher Recovery Using Nitrogen Gas Injection. Energy Summit' 08. Praxair Inc.

Yelling, W.R., Metcalfe, R.S., 1980. Determination and prediction of CO2 minimum miscibility pressures. J. Pet. Technol. 160−168.

Zaks, S.L., 1963. Enhanced Oil Recovery by Gas Injection. Gostoptekhizdat, Moscow.

Zhang, N., Wei, M., Bai, B., 2018. Statistical and analytical review of worldwide CO2 immiscible field applications. Fuel 220, 89−100.

Water altering gas injection 11

Abstract

Gas injection by itself can maintain reservoir pressure and assist in oil recovery. Miscibility of injected gas with oil is very important and injection at high enough for miscibility pressures lead to improved oil recovery. Instability of gas displacement front leads to relatively fast gas breakthrough to the production wells and this is multiplied by low sweep efficiency. Adding displacement agents with high viscosity leads to better oil recovery. Just water can be injected in alternating with gas slugs and significantly improve oil displacement. This is so named WAG technique — Water Altering Gas. The technique can be further modified to include various methods of water and gas mixing and even by adding foam forming surfactants. Good understanding of interaction between reservoir and injected fluids is needed for the maximum process efficiency.

Chapter Outline

The injection of natural or petroleum gas technology has been used to maintain reservoir pressure and increase oil recovery in depleted oil fields. In particular, in the United States it was used much earlier than waterflooding. At the initial stage, gas was injected into the reservoirs at pressures below the pressure of its miscibility with oil. In the period of time when the water flooding into oil reservoirs were not used, applied gas injection technology used to contribute to an increase in oil production and in general oil recovery from horizontal deposits from 5% to 10%. In the inclined formations the technology contributed to between 15 and 25% extraction increase. It was an economically preferred choice as compared to the situations when injection pressure was high enough to dissolve the injected gas.

Later, when the process of artificial flooding in hollow oil reservoirs was widely introduced, it was found that water injection contributes to a greater displacement of oil than the injection of gases that are not miscible with oil. The low efficiency of gas injection as a displacing agent, due to its breakthrough in highly permeable areas to production wells. This decrease in oil production from wells and the overall

Primer on Enhanced Oil Recovery. DOI: https://doi.org/10.1016/B978-0-12-817632-0.00022-0

sweep efficiency of the formation, is associated with a small viscosity of gas compared to water, which produces high mobility ratio.

In order to eliminate the mentioned negative results, combined methods were developed, including the injection of gas and water, so named water alternating gas (WAG) technology. It is believed that in 1951, the gas injection technology was used in conjunction with the contour-boundary flooding to intensify the process of reservoir pressure recovery and increase oil recovery in the Starogroznenskoye field (Russia). The method included sequential with gas injection of water in volumes $0.5-1.0$ m^3 into the gas injection wells. This significantly reduced gas breakthrough into production wells occurrences. It was also found that the periodic injection of water (alternating slugs) not only eliminates gas breakthroughs, but also contributes to an increase in oil recovery from the formation in this pilot site. For the first time in the West, WAG technology was implemented in 1957 in Canada at the North Pembina field.

Later, other variants of WAG were developed and optimized. There are technologies such as: sequential (sequence) water and gas injection, joint (simultaneous) injection of water and gas, as well as foam WAG. The later relies on injection of mixture when a foaming surfactant is added to the injected water.

11.1   WAG methods

The essence of the water alternating gas (WAG) process is the simultaneous or separate injection of water and gas into the oil reservoir.

The gas pumped into the reservoir, which is a non-wetting phase, moves into large pores and into top layers of the formation under the action of gravitational forces. The water on the contrary, under the influence of capillary forces occupies small pores of the hydrophilic reservoir and generally concentrates at the lower strata. Taking into account the uniqueness of water and gas, and in order to achieve a greater effect from the overall injection process, it is advisable to perform alternate, cyclic injection. With this technology, it is necessary that the optimum ratio of the injected volumes of water and gas be proportional to the ratio of the volumes of small and large reservoir pores. The use of water—gas mixtures in this sequence will give the maximum positive effect as a result of combining the properties of the injected water and gas.

Enhanced oil recovery during alternate water and gas injection is also occurs due to the fact that the phase permeability of the wetting phase depends only on the saturation of the formation with water. The increase in oil displacement from the reservoir in the presence of free gas occurs by the value of the maximum gas saturation and is $10-15\%$, at which the gas is stationary.

There are various WAG implementations which are used and differ in the method of supplying the working agents to the reservoir and in they are applied by taking into the account the mixability of the injected gas in the reservoir oil.

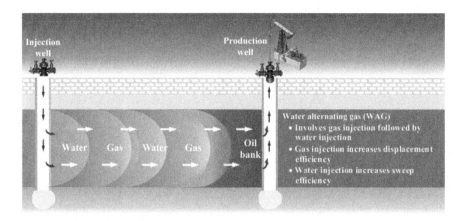

Figure 11.1 Water alternating gas (WAG) process.

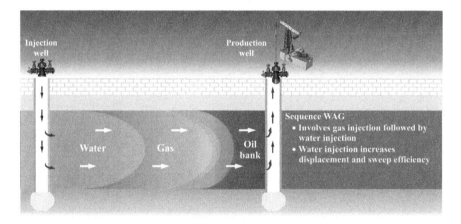

Figure 11.2 Sequential injection of gas and water.

We can list various applications which differ in the way of injected materials:

a. The alternating supply of working agents (see Fig. 11.1) includes gas injection followed by water injection. In this case, gas injection increases the efficiency of displacement, and water increases the reservoir sweep efficiency;

b. Sequential supply of working agents (see Fig. 11.2), includes gas injection followed by water injection. At the same time, water injection increases both the displacement efficiency and the coverage of the formation by the impact;

c. Hybrid WAG (see Fig. 11.3) consists of big injection of gas followed by big injection of water aimed at increasing formation sweep efficiency

d. Simultaneous WAG, when water and gas injected together after mixing at the surface, see Fig. 11.4.

e. Selective simultaneous injection of working agents (see Fig. 11.5). In this case, the working agents are fed into the well through various pipes and mixing is achieved in the oil strata.

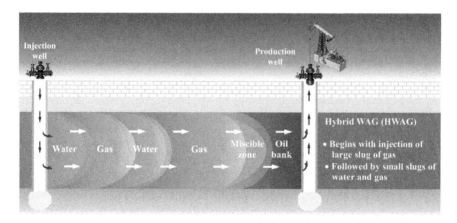

Figure 11.3 Hybrid WAG.

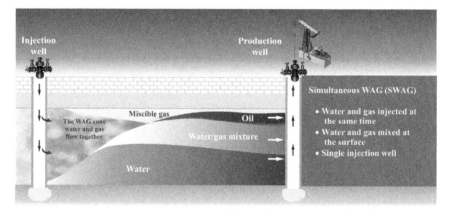

Figure 11.4 Simultaneous WAG.

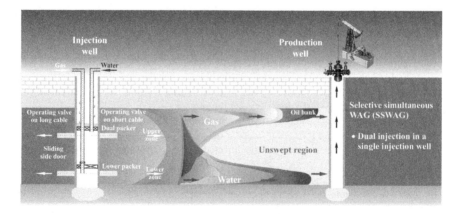

Figure 11.5 Selective simultaneous WAG.

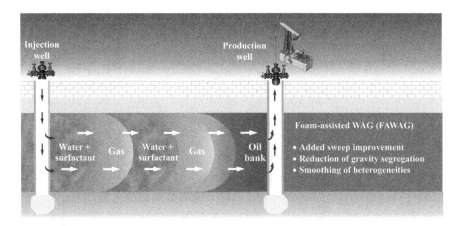

Figure 11.6 Foam-assisted WAG.

f. Foam-assisted WAG (see Fig. 11.6), is carried out by adding a foaming surfactant to the injected water. This provides the best possible reservoir coverage (see Fig. 10.6). At the same time, the gravitational separation of working agents and the influence of rock heterogeneity are significantly reduced.

Depending on just gas flooding or different WAG methods sweep efficiency in the oil containing strata changes as shown in Fig. 11.7. Single (no-water) gas injection leads to fast development of fingering and fast breakthrough. There is also high gravity segregation. Both of those unwanted processes lead to significantly reduced sweep efficiency.

Adding water helps to reduce viscous fingering but still has problems with the vertical gravity segregation. Adding a foam reduces vertical gravity segregation and helps to reduce heterogeneities influence on the sweep propagation through an oil strata.

There are further possible different implementations of WAG depending on the miscibility of gas and strata crude (see Fig. 11.8):

a. WAG in the presence of gas and oil miscibility with pressure above the minimum miscibility pressure. In this case the viscosity of the oil decreases and its mobility increases. Gas dissolves in oil and there is no well-defined front between gas and oil;
b. WAG in the absence of gas and oil miscibility. The process is carried out at a pressure below the minimum miscibility pressure. In this oil as a phase is not affected by the gas and there is a border between the oil and gas phases. The process continues until gas breaks into the production well.

11.2 WAG applicability criteria

As a result of alternating water and gas injection, the sweep efficiency increases by the more uniform interaction with the heterogeneous zones. There is a certain

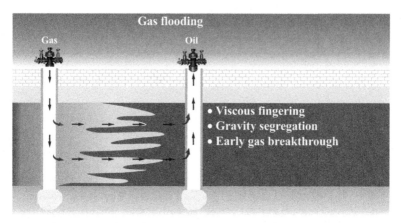

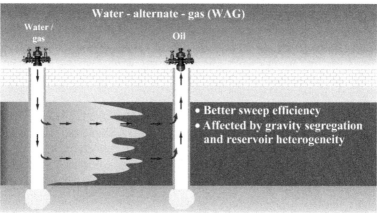

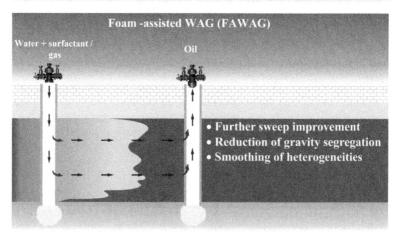

Figure 11.7 Comparison of main processes during simple gas flooding and various WAG implementations.

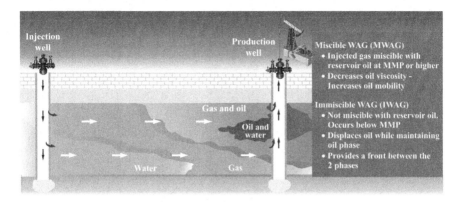

Figure 11.8 Effects of gas/oil miscibility properties during WAG implementation.

blockage of highly permeable volumes with a water–gas mixture and better connection to the zones with low permeability. It has been established that the final oil recovery of a heterogeneous formation, when using any technology of joint injection of water and gas, is higher than when oil is displaced only by water or gas. WAG under optimal conditions increases oil recovery by 7–15% compared to the conventional water flooding. The best effect of WAG application is achieved if the injected gas is evenly distributed over the waterflood reservoir and the water and gas breakthrough to the production wells at the same time. The latter is difficult to achieve and is not always possible. Consequently, in practice, the efficiency of the resulting process is lowered. The high efficiency is impossible to achieve in highly heterogeneous formations.

General screening parameters for WAG can be summarized as shown on Fig. 11.9.

The injection of alternating slugs of water and gas into the formation or simultaneous gas-water mixture through the same injection well also reduces the obtained oil displacement effect. This is due to the fact that after the first cycle of injection of working agents, the phase permeability of the around the injection well decreases, and as a result, the injectivity sharply decreases. This effect can be as high as reduction of injectivity by 8–10 times for gas, and by 4–5 times for water.

The separation of gas and water in reservoir conditions under the influence of gravitational forces produces negative effect the displacement of oil and the coverage of the reservoir by the method. Depending on the heterogeneity of the reservoir and the ratio of viscosities of reservoir fluids, the efficiency of the process can be reduced by 10–20%.

Business considerations come into the effect. For instance, for alternating injection of water and gas, the equipment of each injection well is significantly more complicated.

In order to eliminate the above drawbacks careful process understanding and planning are needed. Injection pressure and slugs volumes should be carefully determined. A significant increase in the number of injection wells with

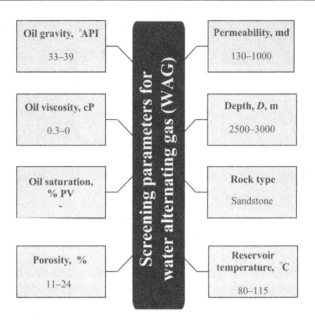

Figure 11.9 Formation parameters for WAG implementation.

corresponding complicated wellhead equipment is necessary. Reduction of capital investment can be achieved by applying acceptable technological and technical solutions. It is necessary to provide the ability to control the injected agents flow and employ benefits of gravitational effects through optimal well locations and well operation parameters.

11.3 Implemented projects

In the completed WAG technology projects air, hydrocarbon gas, nitrogen and carbon dioxide are used as gas. Since the 1950s to the present, about 100 WAG projects have been implemented, with the vast majority in the United States, less than 10% of them were unsuccessful. Fig. 11.10 shows the implemented WAG technology projects.

As can be seen from the figure, WAG technologies with miscible gas injection are most effective. The maximum additional oil recovery was obtained at the Slaughter Estate and Wasson fields where the oil recovery increased by 14.9%.

Fig. 11.11 shows results of WAG applications with hydrocarbons when the injected agents were delivered to the oil strata by different methods.

It can be seen from the figure that regardless of the method of supplying injected agents, on average, fairly high additional oil recovery factors (2.5−19%) were obtained.

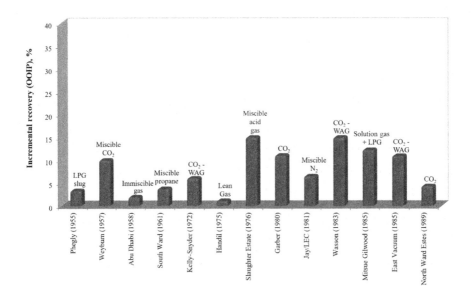

Figure 11.10 Implemented WAG projects.

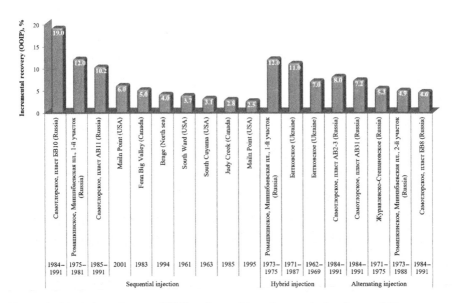

Figure 11.11 Hydrocarbon gas WAG projects with various methods of agent delivery.

Fig. 11.12 shows the results of the implemented WAG projects (using carbon dioxide) with different miscibility of gas and reservoir oil.

The efficiency of the miscible WAG, as it is evident, is significantly higher than that of the immiscible one.

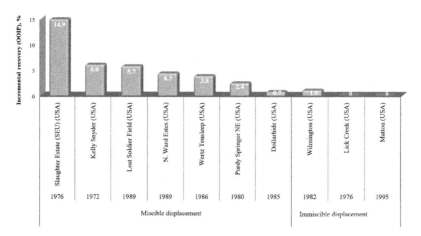

Figure 11.12 Carbon dioxide WAG projects with miscible and immiscible displacements.

Foam WAG is mainly implemented in the North Sea at the fields Oseberg, Beryl, Snorre, Brage. The most famous foam WAG implementation project was carried out at the Snorre field. During the two years from January 1999 to January 2001, 0.8 million m^3 of oil was additionally produced at this field. The total oil production there as a result of the WAG implementation had increased by 22%.

11.4   Technology of implementation

In the majority of cases WAG technology have been applied in the fields which had been already under development for a long time and after primary and secondary oil recovery has already been implemented.

The gases used in WAG projects can be divided into three groups: carbon dioxide, other than hydrocarbon gases and hydrocarbon gases. CO_2 is relatively expensive gas and is commonly used with miscible WAGs. It should be noted that when using CO_2, corrosion has a significant impact on process efficiency and final costs. Hydrocarbon gases are produced in the process of crude recovery and for this reason, especially in most WAG implementations in offshore fields, they are used for WAG. Least of all WAG projects use other non-hydrocarbon gases, in particular nitrogen or flue gases.

Injection system the most commonly used is in five-point configuration with a fairly close distance between the wells.

Tapering or reducing the water−gas ratio is widely employed in the bid to prevent gas breakthrough to production wells. At a late stage of the WAG process

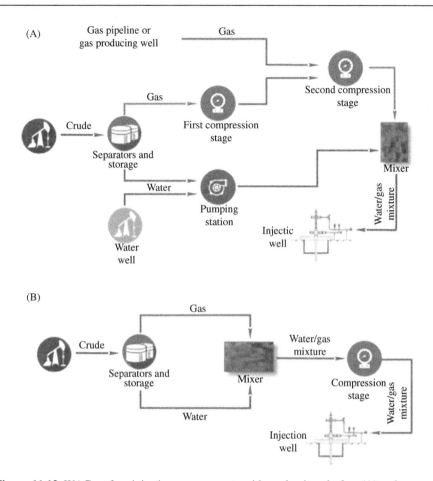

Figure 11.13 WAG surface injection arrangements with predominantly free (A) and associated (B) gas.

the gas slugs are reduced in volume while water slugs are increased. This technique is much preferred especially in the case of expensive gas.

In various WAG variants, various schemes for supplying working agents to the well and wellhead piping are used. In Fig. 11.13 shows possible schemes for the supply of hydrocarbon gas and water to the well during their joint injection, as well as the binding of the wellhead. As can be seen from the figure, both associated and free gas from gas wells can be used in the WAG process.

Injection well for WAG implementation can have a more complex than usual injection structure as shown on Fig. 11.14, when water and gas injected at different depth.

In the case of foam WAG additional equipment should be implemented as shown on Fig. 11.15.

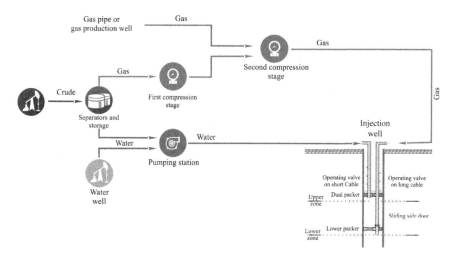

Figure 11.14 WAG hydrocarbons implementation with a separate gas/water injection.

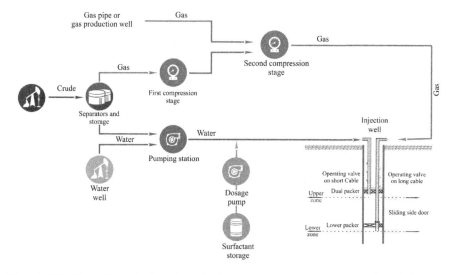

Figure 11.15 Foam WAG hydrocarbons implementation with a separate gas/foam injection.

Further reading

Akhmadeyshin, I.A., 2014. On the technological schemes WAG with simultaneous injection gas and water. Oil Ind. (6), 104–105.

Alvarez C., Manrique E., Alvarado V., Samon A., Surguchev L., Eilertsen T. WAG pilot at VLE field and IOR opportunities for mature fields at Maracaibo Lake. In: Paper SPE-72099-MS Presented at the SPE Asia Pacific Improved Oil Recovery Conference, Kuala Lumpur, Malaysia, 6–9 October 2001.

Chambers, F.T., 1968. Tertiary oil recovery combination water-gas miscible flood − Hibberd Pool. Prod. Mon. 32 (1), 17−119.

Christensen, J.R., Stenby, E.H., Skauge, A. review of WAG field experience. In: Paper SPE-39883-MS Presented at the International Petroleum Conference and Exhibition of Mexico, Villahermosa, Mexico, 3−5 March 1998.

Christensen, J.R., Stenby, E.H., Skauge, A., 2001. Review of WAG field experience. In: Paper SPE-71203-PA SPE Reservoir Evaluation & Engineering., vol. 4 no. 02, pp. 97 106.

Groenenboom, J., Kechut, N.I., Mar-Or, A. Foam-assisted WAG: injection strategies to optimize performance. In: Paper SPE-186991-MS Presented at the SPE/IATMI Asia Pacific Oil & Gas Conference and Exhibition, Jakarta, Indonesia, 17−19 October 2017.

Holtz, M.H. Immiscible water alternating gas (IWAG) EOR: current state of the art. In: Paper SPE-179604-MS Presented at the SPE Improved Oil Recovery Conference, Tulsa, Oklahoma, USA, 11-13 April 2016.

Lo, L.L., McGregor, D.S. WAG pilot design and observation well data analyses for Hassi Berkine South field. In: Paper SPE-84076-MS Presented at the SPE Annual Technical Conference and Exhibition, Denver, Colorado, 5−8 October 2003.

Manrique, E., Padron, R., Surguchev, L., De Mena, J., McKenna, K. VLE WAG injection laboratory field in Maracaibo Lake. In: Paper SPE-65128-MS Presented at the SPE European Petroleum Conference, Paris, France, 24−25 October 2000.

Ning, S.X., McGuire, P.L. Improved oil recovery in under-saturated reservoirs using US-WAG process. In: Paper SPE-89353-MS Presented at the SPE/DOE Symposium on Improved Oil Recovery, Tulsa, Oklahoma, 17−21 April 2004.

Ping, G., Zhonglin, W., Guangtian, T., Kailei, Y., Bin, L., Yukai, L., Maolin, Z. Case analysis on hydrocarbon alternative gas miscible flooding in PuBei oil field. In: Paper SPE-80487-MS Presented at the SPE Asia Pacific Oil and Gas Conference and Exhibition, Jakarta, Indonesia, 9−11 September 2003.

Potter, N., 1996. Water alternating gas injection to lift Miller output by 6%. Offshore 56 (8).

Redman, R.S. Horizontal miscible water alternating gas development of the Alpine field, Alaska. In: Paper SPE-76819-MS Presented at the SPE Western Regional/AAPG Pacific Section Joint Meeting, Anchorage, Alaska, 20−22 May 2002.

Sessions, R.E., 1960. How Atlantic operates the Slaughter flood. Oil Gas J. 58 (27), 91−98.

Skauge, A., Aarra, M.G., Surguchev, L., Martinsen, H.A., Rasmussen, L. Foam-assisted WAG: experience from the Snorre field. In: Paper SPE-75157-MS Presented at the SPE/DOE Improved Oil Recovery Symposium, Tulsa, Oklahoma, 13−17 April 2002.

Sohrabi, M., Danesh, A., Tehrani, D.H. Oil recovery by near-miscible SWAG injection. In: Paper SPE-94073-MS Presented at the SPE Europec/EAGE Annual Conference, Madrid, Spain, 13−16 June 2005.

Spirov, P., Rudyk, S., Khan, A. Foam assisted WAG, Snorre revisit with new foam screening model. In: Paper SPE- 150829-MS Presented at the North Africa Technical Conference and Exhibition, Cairo, Egypt, 20−22 February 2012.

Surguchev, M.L., 1985. Secondary and tertiary EOR methods. M.: Nedra.

Zatsepin, V.V., Maksutov, R.A., 2009. Review of wag process industrial application. Mod. Consist Oilfield Eng. (7), 13−21.

Chemical EOR

12

Abstract

Currently, up to 90% of oil (it is very much region dependant) is produced using water-flooding. The amount of water injected into the reservoir significantly exceeds the level of oil produced. However, the average level of oil recovery coefficient (recovery factor) does not exceed 50%, i.e. up to 50% of oil is left in the formation after waterflooding. This shows the urgency in finding new technologies to increase oil recovery from the reservoirs. Chemical Enhanced Oil Recovery allows to modify water properties, induce chemical reactions in the reservoir, change reservoir fluid properties and transform fluids-rocks interactions in order to increase oil recovery.

Chapter outline

Currently, up to 90% of oil (it is very much region dependant) is produced using waterflooding. The amount of water injected into the reservoir significantly exceeds the level of oil produced. However, the average level of oil recovery coefficient (recovery factor) does not exceed 50%, i.e. up to 50% of oil is left in the formation after waterflooding. This shows the urgency in finding new technologies to increase oil recovery from the reservoirs. The technical ambiguity in the water flooding is that reservoir pressure maintenance requires the injection of large volumes of water, while the injection of a large amount of water leads to the creation of the instability of the displacement front and early breakthrough of water to the production wells. Overcoming this problem is possible with the use of chemical compounds that change the rheological properties of the fluid and surface interaction energies. In some cases both methods are employed. In all cases, if used appropriately, chemical water injection can provide high reservoir pressure and the displacement front stability.

Primer on Enhanced Oil Recovery. DOI: https://doi.org/10.1016/B978-0-12-817632-0.00011-6

Chemical flooding is accomplished by adding one or more chemical compounds to the injected fluid. This leads either to: a decrease in the interfacial tension between the reservoir oil and the injected fluid, or to an increase in the viscosity of the injected fluid. High viscosity reduces the injected fluid mobility. This also leads to increase in the reservoir sweep efficiency.

The main methods of chemical flooding technology are: polymer, micellar-polymer and alkaline flooding. Polymer flooding is used to increase the viscosity of the injected fluid and reduce its mobility. Micellar-polymer and alkaline flooding are used to reduce the interfacial tension between the reservoir oil and the injected fluid. This significantly improves the process of residual oil displacement. Polymer flooding is usually used in the early stages of oil field development with sufficient oil saturation in a porous medium. Micellar-polymer and alkaline waterflooding is usually used at the final stage of field development. The process can be thought as an effective wash of the residual oil

12.1   Polymer flooding

Continuous and effective oil production with high recovery rate is only possible if necessary measures to influence the oil formation by the injection of large volumes of water are implemented. The injection of large water volumes leads to the development of the displacement front unsteadiness and a decrease in the sweep efficiency. The solution to this problem is in creation non-Newtonian, heterogeneous systems, that have adjustable rheological (e.g. flow) properties. All this can be achieved by use of polymeric solutions.

One of the main problems in the flooding of oil reservoirs is the hydrodynamic instability of the oil-water displacement front. The main factor affecting the hydrodynamic instability of the front is the ratio of the viscosities of oil and water in reservoir conditions. Indeed, the critical wavelength of the perturbation λ (water-oil front instability) is inversely proportional to the capillary number C_a:

$$\lambda \sim 1/\sqrt{Ca} \tag{12.1}$$

$$Ca = \frac{v\left(\eta_1 - \eta_2\right)}{\sigma m} \tag{12.2}$$

where: v — speed of the front, η_1, η_2 are viscocities of oil and water, σ — surface tension and m — porosity.

It can be seen from the presented expressions, *ceteris paribus*, that the higher the difference in viscosity of oil and water, the higher the capillary number and the lower the critical perturbation wavelengths. This leads to an early development of the displacement front instability. It is seen as so named process of fingering, when the displacement front is not a straight line anymore. The advance of water continues and leads to the relatively rapid breakthrough process. This means that significant part of infected water channels into the production wells. The reservoirs inhomogeneity also makes water breakthrough to develop much faster as, due to the predominant movement of water in a highly permeable interlayer, layered-inhomogeneous formations greatly reduce the sweep efficiency of the reservoir by the direct waterflooding.

The solution of these problems is based on adding polymers to the injected water, i.e. the use of so named polymer flooding. The mechanism of polymer flooding is determined by three main factors:

1. Increase of injected liquid viscosity (as compared to just water). This reduces the capillary number. When a water-soluble polymer is added to water at a concentration of 0.01−0.1%, the viscosity of the resulting polymer solution increases 3−4 times. This leads to the apparent viscosity of the polymer, during the propagation through the pores, to increases up to 20 times. In this regard, polymer solutions are particularly effective in heterogeneous reservoirs and in fields with high viscosity of oil.
2. Water-soluble polymers used in polymer flooding are by themselves anionic surfactants. Propagation of surfactants in the reservoir will reduce the surface tension. On top of this, in the process of polymer flooding, more surfactant additives are used in the polymer solution. The resulting complex of polymer molecules and surfactants has a molecular weight higher than the molecular weight of the polymer, and hence a higher viscosity. The reduction in interfacial surface tension makes it possible to ensure that residual oil is washed off from the surface of the pore channels behind the displacement front too.
3. Perhaps the greatest influence on the efficiency of polymer flooding is produced by the non-Newtonian properties of the polymer solution. Water-soluble polymers commonly used in oil field practice are characterized by shear thinning, or pseudoplastic rheology (see Fig. 12.1). The plot shows that the faster liquid flow (remember, that the very close to the wall layers almost do not move) the lower is the viscosity. In this case viscosity is low at high speeds of flow and viscosity is at the maximum if the liquid flows very slowly or completely stops.

However, in order to increase the sweep areal efficiency, it is preferable to use injected liquids with shear thickening or dilatant rheology (see Fig. 12.2). The liquid behavior in this case is fully opposite to the described above at some flow speeds. If the liquid flows fast the viscosity increases. This allows to have more

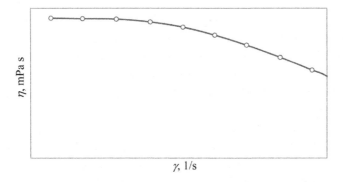

Figure 12.1 Polymer solution shear-thinning (η − viscosity, γ − shear rate).

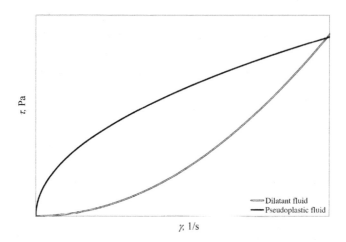

Figure 12.2 Rheology of dilatant and pseudoplastic fluids (τ — shear stress).

stable water-oil bank and reduce speed of infected fluid flow in high permeability channels.

Indeed, assuming that the flow of a polymer solution in a porous medium obeys a power law, the flow rate can be determined by the following formula:

$$v = -B_0 \left|\frac{dP}{dx}\right|^{\frac{1}{n}-1} \frac{dP}{dx}$$

$$B_0 = \frac{n}{3n+1} \, \eta^{-\frac{1}{n}} 2^{\frac{1}{2}\left(3+\frac{1}{n}\right)} m^{-\frac{1}{2}\left(1+\frac{1}{n}\right)} k^{\frac{1}{2}\left(1+\frac{1}{n}\right)} \tag{12.3}$$

For the Newtonian liquid (for the case when viscosity does not depend on shear rate, e.g. no viscosity dependence on the flow speed) $n = 1$, for a dilatant liquid $n > 1$, for the shear-thinning liquid $n < 1$.

Let us consider a simplified flow diagram of the polymer composition in a layered heterogeneous reservoir. Let the filtration of the polymer composition occur in two parallel layers with permeability k_1 and k_2 (with other conditions being the same), and $k_1 \gg k_2$ and corresponds to a power law. Then, according to the above equation, the ration for the flow volumes

$$\frac{Q_1}{Q_2} = \frac{v_1}{v_2} = \left(\frac{k_1}{k_2}\right)^{1+\frac{1-n}{2n}} \tag{12.4}$$

It is immediately obvious that for a dilatant liquid the flow will be much less affected by the permeability variations.

The polymer solution flow in a porous medium is significantly different from the flow defined in the laboratory rheometer. Moreover, diluted polymer solutions which are characterized by shear liquefaction in measurements on the rheometer, can in a porous medium exhibit dilatant rheology, i.e. shear thickening. Such behavior is explained by a retention of polymer molecules by the pore space. At the same time, the polymer macromolecules unfold which increases shear rate and escalate resistance to flow. Retention of the polymer can occur by adsorption or even mechanical processes. In layered-inhomogeneous and micro-inhomogeneous porous media the rheology of the diluted polymer solution is S-shaped (see Fig. 12.3).

At the same time, the dilatant nature of the flow is replaced by pseudoplastic. The change in flow rheology occurs at sufficiently high shear rates and not is observed in the actual oil containing rock. In the formation high-permeable zones higher flow rate is observed compared to the flow rate in low-permeable zones, which leads to an increase in viscosity (decrease in solution mobility). This leads to the flattening of the filtration profile and an increase in the sweep efficiency.

In the course of polymer flow in a porous medium, connate water is also displaced along with the oil. As a result, the polymer solution interacts directly with the reservoir rock, leading to the polymer adsorption from the solution. At the same time, the concentration of the polymer in the solution decreases; a layer of water with a low polymer concentration is formed.

According to various researcher data, the polymer adsorption can reach 150 g/m^3, which is significantly less than the adsorption of surfactants. Usually one of the main requirements for polymers is their minimal adsorption on the surface of a porous medium. However, this is a simplified view of the efficiency of oil displacement with a polymer solution. Adsorption plays an important role in the mechanism of polymer flooding, because it reduces the mobility of the polymer solution (by reducing the permeability) and increase the coverage of the reservoir sweep. At the same time, adsorption should not exceed the limits of the experimentally determined optimal range, since with high adsorption, the front of the polymer movement lags far behind the oil displacement front (the displacement front is formed in this case by water with very low polymer content). This reduces the efficiency of the process due to the fact that oil is displaced mainly by inactive water.

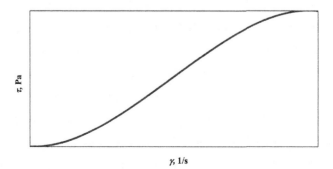

Figure 12.3 Polymer solution rheology for heterogeneous porous media.

Salinity and water hardness have a significant effect on polymer flooding, especially when using hydrolyzed polyacrylamide (PAA, HPAM). The effect of mineralization and water hardness on the polymer solution is expressed in a decrease in the electrostatic repulsion forces in the polymer. Salinity leads to the destruction of the solution structure, a significant decrease in the viscosity of the polymer solution. The effect in some way is proportional to the salt concentration. The overall effect of water hardness has an even more significant effect on the polymer solution, as exposure to bivalent calcium and magnesium ions, other things being equal, is more significant than single-ion ions of sodium and potassium.

Polymer solution viscosity is somewhat proportional to polymer molecular weight. But at the same time big molecules are not very stable and can be broken. This leads to smaller molecular weight polymer solution with a lower viscosity and polymer flooding efficiency reduction. Polymer molecule destruction can be induced mechanically, chemically, thermally and by a bacteria.

Polymers usually delivered in powder form and need to be mixed with water at the injection site. The solution making should include few stages and it is time consuming process. Mechanical mixing needs to be done with care as too vigorous mixing can reduce solution viscosity below stated by the polymer manufacturer. High pressure injection can add to this thinning process. Air oxygen and iron contamination negatively affects polymer molecules. For this reason polymer mixing should be undertaken in the air-tight and preferably iron-free mixers. Bacterial species from the mixed water and anaerobic bacteria in the formation are usually feed on polymer solutions. The reduction of polymer viscosity can be as high as 80%. In many cases this leads to the necessity to use some biocides to suppress bacterial activity.

As a polymer solution is injected its temperature rises to the formation temperature. It is universally accepted that the absolutely top temperature limit for the polymer flooding is always below 130 °C. As the molecule hydrolysis is thermally activated process it is always happening. It is just above some critical temperatures hydrolysis becomes catastrophic. Lately some polymers have been produced which allow to work at slightly higher temperatures.

The injectivity of the polymer solution is an important property for several reasons. Firstly, the rate of injection of the polymer solution directly affects the efficiency of the project. Secondly, cleaning (drainage) of injection wells may be necessary if the polymer solution reduces the injectivity. These cleaning works can reduce the technological and economic efficiency of polymer flooding. Injectivity decreases with increasing molecular weight of the polymer and full analysis and optimization of all processes is highly recommended. The injectivity of the polymer solution is more efficient when the polymer solution exhibits shear liquefaction.

12.1.1   Polymer flooding applicability criteria

The criteria for the polymer flooding applicability can be summarized as on Fig. 12.4.

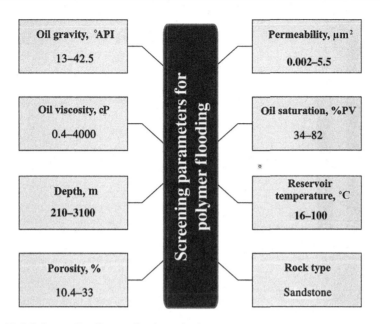

Figure 12.4 Polymer flooding application criteria.

- Oil viscosity should preferably not to exceed 150 cP or be lower than 10 cP
- API gravity should be more than 13.
- The permeability of the rock should preferably be above 10 mD.
- Formation temperature: a low formation temperature is preferred, preferably less than 80 °C and no higher than 100 °C.
- Use wells with high injectivity and with an additional injection pressure boost.
- The clay content in the reservoir should be low.
- Low salinity of injected and formation water is greatly preferred.

Polymer flooding is usually successful in sand formations, but there have been known successful applications in carbonate and fractured reservoirs. It is preferable to use polymer flooding in layered heterogeneous as well as micro-inhomogeneous formations.

It is preferable to use the method in hydrophilic formations, but there are known cases of successful use in hydrophobic and mixed wettability formations.

12.1.2   Polymer flooding completed projects

Starting from the 60s of the last century the active introduction of polymer flooding began As can be seen from Fig. 12.5, currently at around 730 projects have been implemented in 24 countries. At the same time, the absolute leaders here are the United States, where around 560 projects have been implemented, i.e. approximately 76% of all implemented projects.

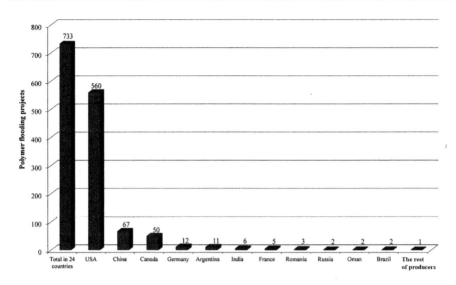

Figure 12.5 Polymer flooding projects in the world.

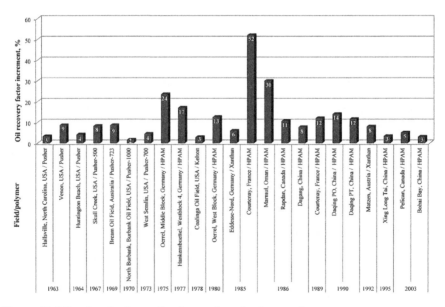

Figure 12.6 Results of polymer flooding projects implementation.

Fig. 12.6 shows successful completed projects and the additional oil produced. As can be seen the use of polymer flooding can significantly increase the oil recovery coefficient.

The vast majority of polymers for flooding are based on acrylamide-based polymers (Pusher, PAA-HPAM) due to their good water solubility, good surface activity and usable rheological properties. Some projects used biopolymer Xanthan (it is produced by bacteria Xanthomonas campestris). The polymer has good stability in salted and high hardness water, it is reasonably resistant to mechanical damage but stable only to approximately 90 °C. On one side it is easily digested by other bacteria, which is not so good for the application, but, on the other side, as it destroyed by the bacteria it is counted as biodegradable and is environmentally friendly.

12.1.3 Implementation Technology

Typically, polymer flooding is carried out with the use of slugs pushed through the reservoir by water (see Fig. 12.7). Depending on the conditions, the size of the slugs may be 10–50% of the pore volume. At the same time polymer dosing is carried out using metering pumps connected to the water supply system in the well.

The first step is to choose the type of polymer. The polymer should:

• ensure the maximum degree of thickening of the injected water and reduce mobility;
• dissolve in water reasonably easily, with no or ultra-low content of insoluble precipitates;
• have a low degree of retention in the filtering process;
• be resistant to shear, chemical, biological and thermal influence;
• have sufficient, economically viable pickup.

The type and concentration of the polymer and the volume of the slug are selected on the basis of the mineral composition of the connate water, the heterogeneity of the reservoir and the pore volume of the site area. It should be taken into the account that when water is highly mineralized (high dissolved mineral content), the concentration of the polymer in the solution should be significantly higher (2–3 times).

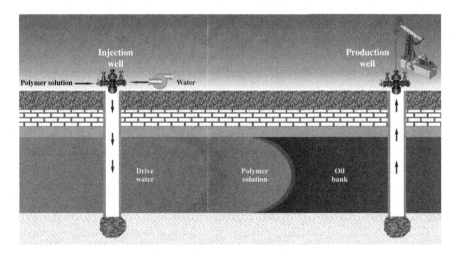

Figure 12.7 Polymer flooding.

For injection of the polymer solution into the reservoir, it is necessary to have an injection pressure much higher than the pressure during normal water flooding in order to provide the necessary economically beneficial filtration rate and the rate of field development. The injection might also need bigger injection well pipes and/or more injection wells. In this regard, the implementation of polymer flooding in reservoirs with low permeability may not be technically feasible.

The placement of wells in polymer flooding does not significantly differ from conventional water flooding if the injection impact is effective and the design parameters of the field development are achieved: injection pressure-volume and oil recovery rates. It should be noted that polymer flooding, usually in-circuit. Some modeling also indicates that combination of horizontal injection and horizontal production wells will be very beneficial.

It is accepted by the EOR application specialists that the polymer flooding is most effective at the early stages of the oil field development. However, an early implementation is usually hampered by the facts that at the early stage there is, usually, not enough information about the oil field geological structure and there are no data on the field injectivity properties even by water. Both those uncertainties usually delay polymer flooding implementation.

12.2 Alkaline flooding

Alkaline flooding includes the use of aqueous solutions of certain chemicals, such as sodium hydroxide, sodium silicate, sodium carbonate and some ammonium compounds. These solutions will react with some components in oil (in general terms the affected compounds usually named saponifiable), in particular with organic acids. The reaction products are surfactants − chemical compounds which significantly reduce the oil/water interfacial tension. Apart from this the oil recovery during alkaline flooding occurs also emulsification of oil and changes in the rock wettability. Some other compounds which can rise pH in the formation are also used if the formation permeability is significantly affected by the sodium presence. It is worth to say that an alkaline flooding only can be effectively applied in specific formations and for specific oils.

Changes in the interfacial properties of oil, water and rock in the process of alkaline flooding have been known for long time. The first patents on alkaline waterflooding appeared in the 20s of the last century, and practical application was carried out at the Baku oil fields in the 40s. Already at the first commercial application a high displacing capacity of alkaline water was found. In subsequent years, interest in the alkaline flooding has increased significantly in connection with the development of oil fields, when oil contains a significant amount of active polar compounds.

Alkaline flooding is based on the interaction of the alkali molecules with organic acids contained in the reservoir oil. As a rule, the emulsifying surfactants are formed in this interaction. The produced surfactant (sometimes called soap) reduces the interfacial tension by two or three orders of magnitude. As shown by experimental studies with an increase in the content of organic acids in oil, the interfacial tension decreases significantly and can be as high as 0.001 mN/m.

The surfactants also increase the wettability of the rock by water. They lead to emulsification of the residual oil. All this results in much more efficient oil displacement from the surface of the rock. Overall, the processes leads to a significant decrease in residual oil saturation and further increase in the oil recovery.

A great influence on the properties of the resulting water-oil emulsion is produced by the organic acids in the oil. The reactions are fairly complex but the overall picture can be produced and simplified by the introduction of an activity of the reservoir oil (as measured by the oil acidity). The acidity is measured by the amount of potassium hydrophyte to neutralize oil acids. All oils then can be, according to the acidity index, conventionally divided into three groups (see Fig. 12.8). When the alkaline solution interacts with the highly active oil there are high surfactants concentrations are produced. The surfactants significantly reduce surface tension in the liquid.

Due to the low interfacial tension, an oil-in-water emulsion is formed. The low-active oils form reverse emulsions − water-in-oil. This emulsion is not stable and this leads to the rheological effect when with an increase in the water content, their viscosity increases. In the case of active oils, on the contrary, with increasing water content, the viscosity of the emulsion is significantly reduced. Regardless of the activity of oil, the excess of an alkali above 0.04% leads to the interfacial tension growth.

An important process during alkaline flooding is change in the rock wettability. This effect is a result of the adsorption of organic acids in the rock from oil. The use of alkaline solutions can significantly reduce the contact angle of wetting of the

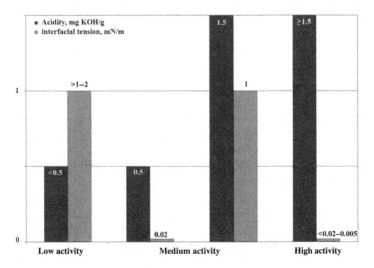

Figure 12.8 Impact of oil activity on interfacial tension (IFT).

reservoir with water, i.e., hydrophilize the rock. In this case, the approaching wetting angle in the sandy rock can be reduced from 70° to 10°. When the wettability of a porous medium changes, two cases should be considered.

- when displacement occurs in a hydrophobic (oil-wet) reservoir, where residual oil is a continuous phase, the alkaline agent changes the pH of the injected water, and the wettability of the rock changes from hydrophobic to hydrophilic. As a result, the coefficient of mobility of the displacement front (injected liquid) decreases, which contributes to an increase in oil recovery.
- even in hydrophilic reservoirs under certain conditions (the reservoir temperature, pH and salinity of the alkaline solution) an intermittent, non-wetting residual oil phase can become a wetting continuous phase. The presence of water droplets in the continuous phase in a hydrophobic formation increases the pressure gradient. The increased pressure gradient allows to displace more oil from smaller pores and the residual oil saturation decreases.

In inhomogeneous formations, emulsification of oil and its capture by small pores, contributes to the deviation of the flow of alkaline solution to areas of the water not covered by the simple waterflood. This significantly increases the areal sweep efficiency and leads to a significant increase in oil recovery.

The presence of salts in the solution significantly changes the process of interaction of oil with an alkaline solution. The presence of calcium chloride increases interfacial tension, reducing the performance of alkaline flooding. The presence of carbon dioxide also leads to reduction the solution activity by to the formation of soda ash, which does not allow to significantly reduce the interfacial tension. Sodium chloride has a positive effect on alkali activity. It allows, other things being equal, to significantly reduce the alkali concentration in the alkaline solution.

The presence of clay in the rock has a significant effect on alkaline flooding. Clays have a negative impact on the process of alkaline flooding. This is mostly produced by the process of the ion exchange. The process leads to a decrease in pH. Alkali adsorption is also different in different rock formations with clays (see Fig. 12.9). As can be seen, on pure quartz sands, the adsorption is absent, but it is very high in montmorillonites and especially anhydrites.

Clays swell in the solution of alkali. Some clays, like montmorillonite, increase their volume more than two folds. Clays also start to produce fines — small clay particles. In effect, the presence of clay significantly affects the efficiency of alkaline flooding. According to the experimental studies, when the content of clays of the montmorillonite group is more than 20%, the anhydrous oil displacement rate with an alkaline solution is the same as with water displacement, while the anhydrous displacement ratio increases due to swelling of the clays, which ensures uniform displacement. In contrast to sand in carbonate reservoirs, the effectiveness of alkaline flooding is associated with the presence of nitrogen-containing compounds in oil. Moreover, the isolation of these compounds on the surface of the carbonate rock contributes to its hydrophilization with all the following processes.

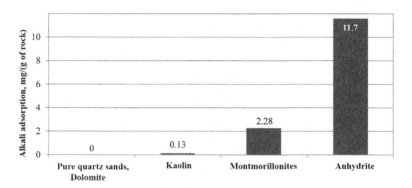

Figure 12.9 Impact of clay type on alkaline adsorption.

12.2.1   Alkaline flooding applicability criteria

Typically, alkaline flooding begins with pre-injection (preflush) of softened water (water with very low mineral content, especially with very low content of calcium and magnesium) followed by injection of the plug of an alkaline solution. The slug volume is at around 10−30%PV. The alkali slug is pushed (driven) through the formation by the injected after the alkali water. In this case, as a rule, alkaline flooding occurs after ordinary flooding, since It is designed to remove residual oil from the surface of the rock due to created low interfacial tension and emulsification of oil. This method is not intended to be carried out at the early stages of field development due to the low potential for sweep efficiency increase (see Fig. 12.10).

The following needs to be taken into the consideration

- Gypsum. Reservoirs with layered anhydrite content of more than 0.1% should be rejected as candidates. This is due to the high adsorption of alkali from the alkaline solution.
- Kaolinite. In fields with a high content of kaolinite, alkaline flooding can be carried out with alkalies with a low pH (8.2−10).
- Montmorillonite. Montmorillonite, due to its high surface area and cation-exchange ability, can absorb most of the alkali introduced due to adverse precipitation reactions. High montmorillonite rich collectors are generally not suitable for alkaline flooding.
- Sandy formations are most preferred for alkaline flooding. Carbonate rocks are less preferred because alkalies with high pH react with carbonates.
- The CO_2 content in the oil reservoir is currently considered as an important selection criterion. Reservoirs with a high content of CO_2, and the presence of formation waters with a pH of less than 6.5 are not good candidates for alkaline flooding.

As noted above, in general, reservoir oil must have a high acid number − to achieve low interfacial tension under alkaline flooding especially in the case when the injected solution does not contain a synthetic surfactant. Nevertheless, even in fields containing low-active oil, alkaline water-flooding can be successful as a result of the manifestation of other, in addition to a decrease in the interfacial tension, oil displacement mechanisms.

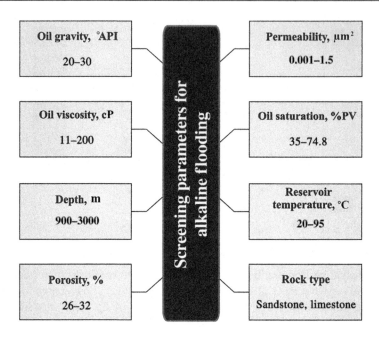

Figure 12.10 Alkaline flooding application criteria.

Indeed, let us consider the process of alkaline solution filtration, with low interfacial tension in a heterogeneous oil reservoir. In this case we assume that the porosity has sets of big and small pores. During water flooding the water is predominantly adsorbed into capillaries of small diameter under the action of capillary pressure. According to the Laplace equation the capillary pressure is associated with interfacial tension:

$$P_k = \frac{2\sigma\cos\theta}{r} \tag{12.5}$$

where P_k is the capillary pressure, in MPa; σ is the interfacial surface tension, in N/m; Θ is wetting angle, in degrees; r is the radius of the capillary, in meters.

As can be seen from formula 12.5, a decrease in σ by the use of alkalies can significantly weaken the absorbing capacity of water in small-diameter capillaries under the action of capillary pressure, which can lead to a decrease in oil recovery of low-permeable interlayers. In highly permeable layers, in which the effect of capillary absorption can be neglected, the alkaline solution, due to the low interfacial tension and emulsification of oil, will increase oil recovery.

It is possible to increase the oil recovery from small-diameter capillaries with the use of water characterized by an increased value of σ. Freshwater, sea water, chlorine-calcium type formation water, connate water solutions have rather high interfacial tension values. However, in the high-permeable layers, the washing properties of such water are much worse than those of an alkaline solution. In order to address all this, an alternate liquid injection with high and low interfacial tension is used. For example, water and alkaline solution (see Fig. 12.10) alterations are used. In this case, just water is first injected and has high absorbent properties, and then slug of liquid with improved washing properties is injected. Water, characterized by a high interfacial tension value, is sucked into the capillaries of small diameter and displaces oil from them into large pores. As a result of the subsequent injection of an alkaline solution having improved washing properties easily moves oil towards production well. This leads to increase in oil recovery from both low and high permeable interlayers. However, this situation is possible only in the case when the flow of water in the capillaries of small diameter was greater or at least equal to the rate of flow of water in the capillaries of large diameter.

Let us estimate the ratio of the flow rates of water in the capillaries of small and large diameter. The rate of movement of water in the capillaries of small diameter can be determined from the following expression:

$$v_k = \frac{R\sigma}{4\mu\rho L} \tag{12.6}$$

where: v_k − liquid speed in small capillary: R − capillary radius; μ − kinematical viscosity; ρ − liquid density; L − capillary lengths.

Liquid speed in the pores of big diameter can be defined from the following equation:

$$v_f = \frac{Q}{Fm} \tag{12.7}$$

where: v_f − liquid speed flow in big pores; F − cross-section area of the oil containing formation; Q − injected water volume; m − porosity.

In order for oil from small-diameter capillaries to be displaced into large-diameter capillaries, the front of capillary impregnation must be ahead or at least keep up with the front of the water flow in large pores, i.e.

$$v_k \geq v_f \tag{12.8}$$

In the oil formations we can expect the following spread of values: $R = 10^{-8} \div 10^{-6}$ m; $\sigma = 10^{-3} \div 10^{-1}$ N/m; $\mu = 2 \cdot 10^{-7} - 2 \cdot 10^{-6}$ m²/s; $\rho = 10^3$ kg/m³; $L = 10 \div 5 \cdot 10^2$ m; $Q/F = 0 \div 10^{-5}$ m/s; $m = 0.2 - 0.35$.

Use of those values in the above equations produces the following values for the liquid speed:

$v_k = 2.5 \cdot 10^{-12} - 1.25 \cdot 10^{-5}$ and $v_f = 0 - 5 \cdot 10^{-5} м/с$.

Consequently, the range of change in the flow rate in large pores completely overlaps the range of change in the rate of flow of water in small-diameter capillaries. This shows that by adjusting the rate of water injection for all possible conditions it is possible to fulfill the condition of 12.8.

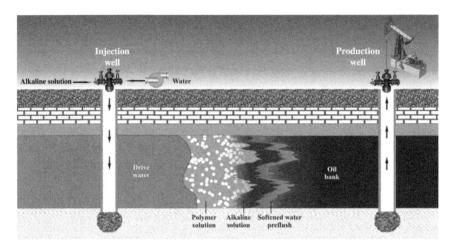

Figure 12.11 Alkaline-polymer-surfactant flooding.

In addition to this, just alkali injection method, there are numerous, more advanced and more expensive options for the implementation of alkaline flooding. One of them is to inject a polymer solution slug with after the alkaline solution to control mobility and improve displacement efficiency (see Fig. 12.11). It is more expensive and technologically more challenging option but it can be viewed as economically justified. Due to the complexity of mineralogy and lithology of oil reservoirs, the probability of a reaction between an alkaline solution, salt water and oil is significant under reservoir high temperature and high pressure conditions. In this regard, the choice of the appropriate injection system for specific oil field conditions is very important for obtaining the best results.

In modern implementations of alkaline flooding, alkaline agents are used in combination with low concentrations of synthetic surfactant and polymer solution slugs to control mobility.

Rather different method of alkaline flooding was first implemented in the former USSR, in 1976, at the Trekhozernoye deposit. This method includes the pre-injection of freshwater, the injection of hard reservoir water (connate water) and alkaline solution. As a result of the sedimentation reaction between the alkaline solution and the hard formation water, the permeability of the highly permeable parts of the formation decreases and the formation coverage of the formation increases. Periodic, cyclic pumping of liquids in the specified sequence is also possible (Fig. 12.12).

Usually for the preparation of alkaline solutions the following chemicals are used:

- caustic soda, sodium hydroxide — $NaOH$;
- soda ash, sodium carbonate — Na_2CO_3;
- ammonia — HH_4OH;
- liquid glass, sodium metasilicate Na_2SiO_3.

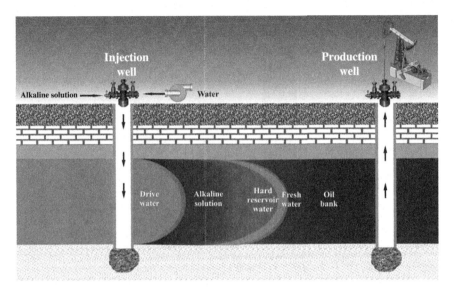

Figure 12.12 Alkaline flooding with precipitate forming solution.

The concentration of the solutions is selected depending on the reservoir, lithological properties of the reservoir and is determined by laboratory tests on core samples using reservoir fluids. The main task of laboratory tests is to determine the concentration of alkali, which ensures the lowest possible interfacial tension. Usually, taking into account adsorption from a solution, the concentration of sodium hydroxide is 0.2−0.4%. However, in hydrophobic formations and with high salinity of formation water, higher alkali concentrations in the solution (up to 5%) may be required in order to modify the wettability of the rock. As a rule, alkaline solutions are pumped in the form of a slug with a volume of 10−30% of the pore volume of the reservoir (exact values are depending on heterogeneity, lithology of the reservoir, adsorption and other alkali losses). The alkali slug is pushed through the reservoir by water. At the same time, to save the alkali, softened water (sodium carbonate solution) water is pre-pumped before the alkaline solution is pumped.

In some implementations of alkaline flooding, a cyclical effect and a change in the direction of flow of the fluid is carried out.

It is known that high-viscosity oils are generally more active when interacting with alkalies than low-viscosity oils. However, the use of conventional alkaline flooding does not significantly improve oil recovery. Therefore, the implementation of thermal alkaline flooding is possible when alkali is added to the injected steam. In this case an areal flooding with a dense grid of wells is preferred.

12.2.2   Alkaline flooding completed projects

Fig. 12.13 demonstrates the results of the first implementations of alkaline flooding in the United States in the 1970s. In all these cases, alkaline flooding was carried

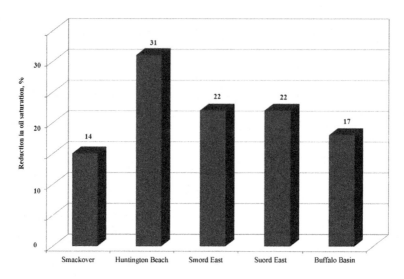

Figure 12.13 Results of alkaline flooding projects implementation.

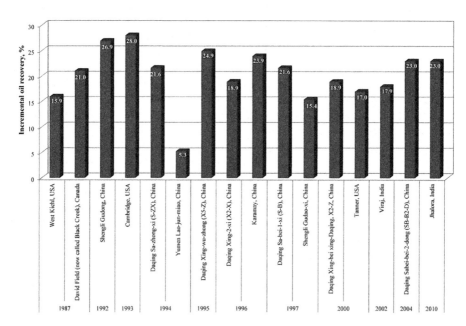

Figure 12.14 Results of alkaline flooding projects implementation.

out after ordinary waterflooding. The residual oil saturation in most cases has changed significantly (up to 31%), which indicates the technological efficiency of the project. It should be noted that in the last century, about 50 projects of alkaline flooding were implemented in the USA alone.

Fig. 12.14 provides published data on majority of pilot and large-scale world-wide from 1992 to 2011 alkali flooding implementations. Information on other combinations of alkaline flooding such as alkali surfactant (AS), alkali polymer (AP), and alkali surfactant polymer foam (ASP-foam) is not given. If these combinations are taken into account, in general, about 100 pilot and large-scale projects have been implemented using alkalies in the world. Many of the implementations were ineffective or with difficult to estimate the effect.

Of the reported 32 successful project, 21 projects were implemented in China, seven in the United States, two in India, one in Canada, and one in Venezuela. Only one project in Venezuela was implemented in marine conditions. As can be seen from the figure, the best result was achieved at the Cambridge, USA field.

In all reported successful projects a five-point or reverse five-point waterflooding system were used.

Microbial enhanced oil recovery (MEOR)

13

Abstract

One of the most promising enhanced oil recovery methods in development and early field applications is the use of microorganisms (microbes). In many ways it can be counted as a method of high adaptability, low capital intensity and relative safety to the environment.

Chapter outline

One of the most promising enhanced oil recovery methods in development and early field applications is the use of microorganisms (microbes). In many ways it can be counted as a method of high adaptability, low capital intensity and relative safety to the environment.

There are at least two main approaches in microbiological methods application for enhanced oil recovery:

— Production of metabolic compounds (bio-surfactants, biopolymers) at the ground level and their injection into the reservoir;
— Establishment and support of microorganisms activity directly in the oil reservoir.

The ability of polysaccharides of microbial origin to change the rheological properties of water, causing the formation of a gel, led to an early increase in interest in these compounds for their use in oil production. The biopolymers are produced at the ground level and can be done at any suitable location. Microorganisms propagation and cultivation is done in special tanks, where all the necessary conditions are created for them. Currently, industrial production of a number of microbial polysaccharides is carried out. Such products as xanthan, sclerglucan, emulsan, ritizan, guran, etc. are produced this way and are widely used in polymer EOR.

Microorganisms are also able to produce and secrete surfactants. Microbial surfactants are often good emulsifiers. Studies have shown that surfactants of microbial origin, due to their physicochemical properties, can be an effective means of enhancing oil recovery.

Primer on Enhanced Oil Recovery. DOI: https://doi.org/10.1016/B978-0-12-817632-0.00012-8

Practice has shown that provision for production of biopolymers and bio-surfactants at the surface level in special bioreactors, mixing the products with water and injection into the formation is relatively costly and requires significant investments. However, this approach has significant advantages. Due to high quality, effectiveness and stability of the bio-products, the bio-products can be applied in the oil deposits with a relatively high reservoir temperatures and highly mineralized waters.

The method of microbiological impact on the reservoir *in situ* is based on an oil displacement rate enhancement. This enhancement is a result of the intensification of living activity of microorganisms capable of digesting and converting complex organic compounds. The digestion products are more chemically simple compounds. The secretion products of microorganisms are liquids (fatty acids, alcohols, solvents) and gases (CO_2, CH_4, N_2, H_2). All those metabolic products are good oil properties modifiers and oil displacers. Certain groups of microorganisms are able to form biopolymers and bio-surfactants within an oil reservoirs.

The mechanism of microbiological activity in a reservoir is complex and it is a multistage process. Firstly, oxygen or air with high amount of easily digestible for microbes materials are injected into the formation. In the reservoir after a nutrient flooding, decomposition and oxidation of hydrocarbons injected by microorganisms proceeds in the presence of oxygen. This results in the microbes secretion of fatty, naphthenic and aromatic acids, alcohols, ethers and other solvents. The increase in the biomass of microorganisms happens due to the correctly selected nutrient medium.

A characteristic feature of oil, as it has been mentioned so far many times, is its exceptional heterogeneity. Oil consists of organic compounds, markedly different in chemical composition, structure and properties. Individual fractions of oil in variable degrees are subjected to the microbial digestion and decomposition.

Secondly, oxidation products of petroleum hydrocarbons, due to oxygen injection, become food substrates for microorganisms deep in the formation, where oxygen is absent. Produced fatty acids and alcohols are used by other microorganisms types. The later stage produces acetic acid, carbon dioxide and hydrogen. In the final phase, methane is formed from the intermediate decomposition products, which is the final link in the chain of transformations of organic substances in the formation. This demonstrates complex chain of processes − at the beginning there is a decomposition, consumption of more easily digestible organic compounds, then a consumption of compounds more difficult to utilize. All this leads to the situation when in the reservoir there are mixtures of all processes products from every stage.

The scheme of oil displacement during microbiological processes is shown in Fig. 13.1.

The microbial cells activity is supported by the presence of enzymes − organic compounds that catalyse certain reactions. Organic substances, introduced by the culture liquid (oxygen plus nutrients plus bacteria) into an oil reservoir, undergo a chain of transformations under the influence of the functional activity of all, injected and reservoir present, microorganisms.

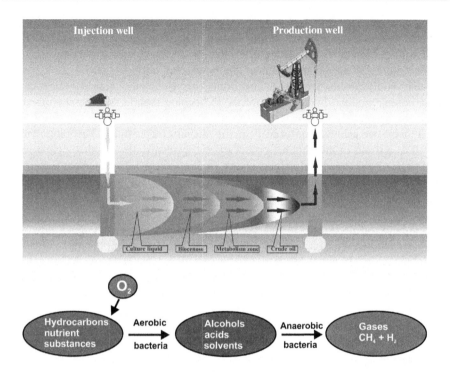

Figure 13.1 Bacteria culture liquid injection and oil permutations for microbial enhanced oil recovery.

Thus, the following processes occur, as in stages from injection to production well, during systemic nutrient flooding of an oil reservoir:

- as a result of aerobic hydrocarbon-oxidizing microorganisms activity intermediate oxidation products are formed. Those including acetate (CH_3COO^-) and carbon dioxide (CO_2),
- further from the injection well, in the anaerobic zone, as a result of the activities of other groups of microorganisms — fermenters,
- at the last stage, hydrogen and methane are released.

In microbiological processes evolution, the products of one bacteria population serve as a nutrition for the next population. In this regard, the compounds formed directly in different zones of the reservoir, have a significant impact on the physico-chemical properties of all constituent in the oil-water-porous rock system. Acids and alcohols, for example, dissolve carbonate rocks, prevent scaling and thus contributes to an increase in porosity and higher rock permeability. Solvents are directly involved in the oil thinning and small droplets extraction from the pores. Moreover, solvents also change rock wettability.

During digestion of oil by microorganisms some gases are released. The gases have a multifactorial effect on reservoir fluids: increase pressure, change pH and viscosity of water, compressibility of oil and saturation pressure. The changes go on: oil density, viscosity and volume expansion are modified, surface tension on oil/water interface are altered. All this usually positively affects phase permeability through formation. Rather negative effect can be observed in clay swelling.

The microbial surfactants reduce the interfacial surface tension at water/oil boundary and at the rock surface. This allows for globules of residual oil, retained in the porous medium by capillary forces, to become mobile. Microbial surfactants have a high emulsifying ability and form highly dispersed oil-in-water emulsions, which helps to achieve uniformity in the displacement front. The biopolymers formed as a result of the vital activity of microorganisms alter the rheological properties of reservoir fluids and help to reduce fingering in the displacement front.

The above processes in their synergy have a significant impact on the final oil recovery. It is obvious that the microorganisms mechanism of enhanced oil recovery has the same base as in other EOR techniques. However, it is difficult to isolate the dominant oil displacement mechanism in this case. All of the above processes take place simultaneously. It is obvious and highly probable that the high efficiency of biotechnology is not a function of any one single factor, as is the case when using physicochemical methods of reservoir stimulation. The effect of MEOR is determined by the integral synergistic effect of many processes.

One or another process or mechanism at each individual physical (formation) site or even strata may play a greater or lesser role in the oil displacement process. One can try to map bio-products and bio processes to the basic reservoir stimulation techniques as it is shown at Fig. 13.2. Realization of prevailing impact mechanism nevertheless depends on the formation geological conditions.

Experimental studies on flat horizontal reservoir models revealed that microbiological exposure contributes to an increase in the oil displacement ratio by 15−20% compared with ordinary water.

When using the microbiological method of enhanced oil recovery, based on the functional activity of hydrocarbon-oxidizing microorganisms in the fields producing paraffinic oil, it is possible not only to significantly increase the oil recovery in the

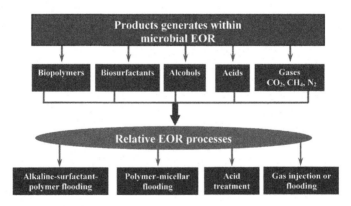

Figure 13.2 Relation between microbial EOR products and basic reservoir stimulation processes.

strata, but also to prevent the deposition of solid hydrocarbons. For this process to work, the provision of the bioactive zone with air or oxygen is imperative. Microorganisms are able to use hydrocarbons of different classes of simple and complex structure. Almost all hydrocarbons that make up oil can be subjected to microbial digestion. Nevertheless the digestion speed depends on the exact chemistry and even on the phase state. For instance, microorganisms multiply and process nutrients most quickly and effectively on solid n-paraffins while the processes are somewhat slower in liquid and gaseous n-paraffins.

It is known that increasing the concentration of carbon dioxide in the gas mixture will help to reduce the amount of precipitation of calcium and magnesium salts. The increase in carbon dioxide content in the oil reservoir can be achieved by nutrient flooding of the reservoir with a bioreagent containing carbohydrate compounds which is decomposed by microorganisms during fermentation to CO_2 For this purpose molasses or whey can be used.

The use of biotechnologies associated with the selective action of microorganisms can be beneficial to improve the efficiency of waterflooding of oil fields and to increase the sweep efficiency in the reservoirs with very high heterogeneity. Microorganisms quickly clog the high permeability zones, where a large amount of injected fluid enters. The expected efficiency and selectivity of blocking highly permeable interlayers of biomass of microorganisms is the highest when both microorganisms and nutrients are injected into the oil reservoir. At the same time, the biopolymers, formed as a result of nutrient digestion by the microorganisms, increase viscosity of the formation water. This increases oil displacement by advancing finger-free water front.

In some cases, waterflooding of oil reservoirs may be accompanied by functionality activation of sulfate-reducing bacteria. In this case, in nutrient containing water injection it is necessary to focus on the factors limiting the activity of sulfate-reducing bacteria (SRB).

The main problems that can be encountered in the implementation of microbiological methods, in addition to sulfate reduction processes, are related to the transport of metabolic products into the reservoir, ensuring the necessary concentration of microorganisms for their growth and development, as well as the possibility of optimizing the required microbiological activity.

13.1   Applicability criteria

The geological and physical conditions of oil reservoir considered for the microbiological treatment must meet certain requirements for the temperature, pressure, salinity and mineralogical composition of connate water. Biotechnologies can be applied most effectively at fields with terrigenous and sandy oil reservoirs. The formation should have sufficiently high porosity (more than 20%) and reservoir permeability (more than 0.04 μm^2). The oil strata temperature should be less than 100 °C. All criteria are summarized on Fig. 13.3 Highly mineralized connate water

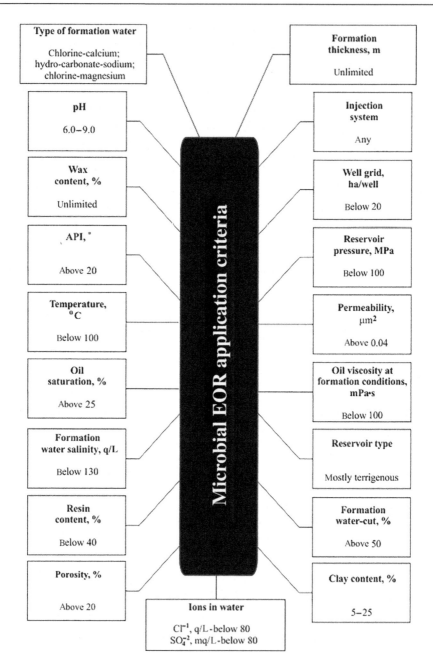

Figure 13.3 Microbial EOR application criteria.

suppresses the development of microbiological activity and processes. The overall salinity of water in the reservoir should not exceed 130 g/l, and the content of sodium chloride below 5−10%. The content of SO4 in the reservoir and injected waters should not exceed 80 mg/l in order to prevent the development of sulfate reduction. The oil reservoir must be sufficiently washed to achieve a large area of water-oil contact.

13.2 Implementation technology

Most of the developed microbiological methods of enhanced oil recovery are based on the introduction into the reservoir as bioreagent production waste from other bioproduction applications. In this regard, the capital costs of the process consist of the cost of shipping and injection of materials. The implementation process of microbiological stimulation does not require any special equipment. Usually, the volume of the injected liquids must be at least 0.1% of the pore volume in the area of implementation. After the bioreagent is injected, the wellhead is sealed for an incubation period (1−2 weeks), then the area is put into operation, e.g. oil production.

Control over the process is carried out using field research methods of the oil volume and composition in the production wells and also by hydrodynamic studies and geophysical methods. In order for the microbiological processes in the reservoir to not slow down it is necessary inject periodically microorganisms and nutrient substances into the formation. The cycles of nutrient flooding are determined on the basis of laboratory and field studies.

To improve microbiological methods of enhanced oil recovery and obtain an integrated effect, it is advisable to combine them with other reservoir stimulation technologies. It is well known that hydrocarbons in oil are fully reduced organic compounds and can be oxidized by microorganisms mainly in the presence of free oxygen, i.e. in aerobic conditions. In the anaerobic zone, in the absence of oxygen, certain groups of fermenting microorganisms consume carbon dioxide, forming hydrogen and methane. With an increase in the concentration of oxygen and carbon dioxide, their metabolism rate increases to a certain maximum peak value. With a further increase in the oxygen concentration, their metabolism rate does not change significantly. Combining microbiological and gas formation stimulation methods (air and/or CO_2 injections) or their alternation, a synergistic effect can be obtained. For a detailed study of the processes, identifying the dominant mechanisms of stimulation integrated methods, additional, field related, the reservoir focused research is strongly recommended. The greatest efficiency is achieved in cases where, when combining methods, their mutual influence on the oil immobilization mechanism is synergetic and well balanced.

The source of energy for microbial cells, as it is well known, are the components of organic polymers and surfactants. Sufficient availability of nutrients significantly accelerate the processes of cell colonies growth and it increases volume of metabolic products. At the same time, for the success of biochemical decomposition reactions, the presence of compounds of trace elements such as K, Na, Ca, Mg, etc. is vitally important. Those elements and their compounds are usually present in the reservoir connate water. It is vitally important that all this is well understood and taken into the account when other physicochemical methods of reservoir stimulation are used. The best outcome, highest oil extraction, are achieved when all methods, such as injection of polymer solutions, surfactants, alkalies, mycelial solutions in combination with bio-effects are used. Even higher results are achieved at the well calculated and executed periodic sequential stimulation with various solutions for the full synergistic effect.

It should be stressed again that microbiological methods of enhanced oil recovery as compared with traditional physicochemical reservoir stimulation methods are low-cost, are simple in implementation and can be applied in deposits with hard-to-recover oil reserves. Biotechnologies not only contribute to a significant increase in oil recovery, but also reduce the impact of water influx on operational performance. Along with this, microbiological methods allow solving the issues of creating waste-free technologies and improving the environmental situation of the developed fields. This all provides solid foundation for the further MEOR development and expansion of MEOR applications.

13.3   Implemented projects

The first commercial microbiological effects on the reservoir began in the middle of the last century. One of the main factors ensuring the effectiveness of biotechnological methods is the correctly chosen nutrient substrates and the environment for the microorganisms. Microbiological effects on the reservoir can be divided into methods based on the activation of aerobic processes and methods based on the activation of anaerobic processes. However, aerobic methods simultaneously stimulate the development of anaerobic and vice versa. For this reason, there can be no purely aerobic or anaerobic processes, there are processes in which this or that type of processes prevails.

Currently, the following technologies are being developed, based on the formation of metabolic products directly in the reservoir:

- a culture of microorganisms is introduced into the formation together with a nutrient substrate;
- only a nutrient substrate is introduced into the reservoir to activate the native reservoir microflora.

The most widespread technology is "molasses fermentation". It is based on the introduction into the reservoir of molasses and microorganisms that can ferment hydrocarbon substances. The result of microbial activity in this case is production of bio-methane and CO_2 In this case the preference is given to the injection of mixed cultures of aerobic and anaerobic microorganisms. Various materials, such as yeast brew, whey, aerated solutions of nitrides and phosphorus salts, are nutrients in the injected mixtures. This technology, and indeed the technology for ensuring the activity of microorganisms directly in the reservoir. This method has become widespread in the fields of Azerbaijan, where it has been introduced since the 1980s. The technology was introduced at seven well in the developed oil fields. The average technological effect (output—input ratio) was 7.2 tons of oil per 1 ton of injected mixture. In total about 200 thousand tons of additional oil was produced.

The implementation of MEOR technologies, based both on the production of metabolic products on the surface level and injecting them into the reservoir, and on ensuring the activity of microorganisms directly in the reservoir were carried out in the following projects:

- use of MEOR in the Daqing oil field has allowed an increase in oil production by 11.2%, while reducing the oil viscosity by 38.5%.
- Portwood J.T. presented performance analysis and 322 MEOR projects conducted in more than 2000 oil producing wells in the United States that used the same MEOR technology. The analysis showed that in 78% of cases there was a positive effect. The oil production increased in averaged by 36%. It is concluded that MEOR is environmentally friendly. From the general operational point of view, problems associated with solid paraffin accumulation, emulsion formation and metal corrosion were significantly reduced. This MEOR technology has demonstrated cost effectiveness and high feasibility. The average return on investment was 5 to 1 during the first 24 months of MEOR, and the average payback period of the project was six months.
- Successful implementation of MEOR was carried out at the Daqanq field. Over three years of use in 77% of producing wells located in the implementation zone, an increase in oil production with technological effect (input—output ratio, injected mixture to produced oil) of 1:5.2 was observed.
- In Southern Alberta (Canada), at the field X had a MEOR injection that lasted for a year. The average monthly production from one well from 21 BOPDs in July increased to 51 BOPDs in September 2014. During September 2014, about 14,000 bbl of oil was additionally produced.

Further reading

Guo, H., Li, Y., Yiran, Zh., Wang, F., Wang, Y., Yu, Zh., et al. Progress of microbial enhanced oil recovery in China. In: Paper SPE 174697 MS Presented at the SPE Asia Pacific Enhanced Oil Recovery Conference, Kuala Lumpur, Malaysia, 11—13 August 2015.

Havemann, G.D., Clement, B.G., et al., 2015. New microbial method shows promise in EOR. JPT. March.

Ibragimov, Kh.M., Abdullayeva, F., Guseynova, N.I. Experience of microbial enhanced oil recovery methods at Azerbaijan fields. In: Paper SPE-177377-MS Presented at the SPE Annual Caspian Technical Conference & Exhibition, Baku, Azerbaijan, 4—6 November 2015.

Li, Q., Kang, C., Wang, H., Liu, C., Zhang, C., 2002. Application of microbial enhanced oil recovery technique to Daqing oilfield. Biochem. Eng. J. (11), 197—199.

Portwood, J.T. A commercial microbial enhanced oil recovery technology: evaluation of 322 projects. In: Paper SPE-29518-MS Presented at the SPE Production Operations Symposium, Oklahoma City, Oklahoma, 2—4 April 1995.

Forefront EOR

14

Abstract

Many new technologies are developed for the Enhanced Oil Recovery. Newcomers to the field are at the different stages of applicability, complexity and technical readiness level. Before considering any of them it is the best to understand well local circumstances, resource availability of all sorts and the reservoir itself. There are big reserves of heavy oil worldwide and development of Enhanced Oil extraction for them became one of the main priorities for EOR techniques.

Chapter Outline

Many new technologies are developed for the Enhanced Oil Recovery. Newcomers to the field are at the different stages of applicability, complexity and technical readiness level. Before considering any of them it is the best to understand well local circumstances, resource availability of all sorts and the reservoir itself. There are big reserves of heavy oil worldwide and development of Enhanced Oil extraction for them became one of the main priorities for EOR techniques.

We will start with a Low salinity water injection, which in a way has been used for a while but only now is getting better understanding, broader development and the community appreciation. The advance of other techniques will greatly depend on their application robustness, state of the economy and political will.

Primer on Enhanced Oil Recovery. DOI: https://doi.org/10.1016/B978-0-12-817632-0.00014-1

14.1   Low salinity water injection

Let us say again that the water in the reservoir (connate water) can be divided into: irreducible water (water on the stone surface), free water and clay bonded water. It is possible to say that almost in every extracted liquid from the well we would have some amount of water. It makes good economic sense during oil extraction to separate the hydrocarbons from the extracted water at the extraction site. Separators are used with great susses for this purpose. After water separation from the oil one needs to deal with a relatively large amounts of water and somehow dispose it. The separated water, sometimes so named connate water, has very high mineral and chemical content. The untreated separated from hydrocarbons connate water is very corrosive to the site metal infrastructures and it is also health and environment hazard if left without the treatment and/or purification. There are strict regulations, country dependant, on this water disposal.

It was realised more than a century ago that the separated connate water can be pumped down, re-injected, into the formation. This solves the problem with the water disposal and at the same time acts as a method for partial reservoir pressure maintenance. We only use word "partial" as for the full pressure maintenance one will need more water or to use some gas too. For the last 90 years this re-injection is widely used in the oil industry.

Generally, water injection is most widely used secondary oil recovery technique. In many cases this is due to water availability, injection simplicity and low cost and sufficient efficiency with which water displaces oil. Additionally used, to the connate, water need to be compatible with the connate water and the reservoir chemistry. The injected water should be cleaned from bacteria, oxygen and preferably from many divalent ions. The amount of injected water during waterflooding is always significantly bigger than the volume of produced oil. This is partially due to the fact that significant amounts of injected water ends in the production liquids. The water percentage in the production fluids increases over time as water inevitably finds a way into the production wells. Injected clean, well formulated water (brine) ends as a complicated cocktail of different dissolved compounds which in some cases contains even radioactive elements.

Overtime it was realised that the injected water in many cases does much more than just to mechanically push oil towards production wells. In essence, water by itself is a chemical agent and alters chemistry of the reservoir. Chemical balance in the formation shifts and this alters many pre-existing balances. As the result, we must account for a whole chain of alterations in reservoir equilibria. This re-balancing proceeds in time and reservoir volume.

Regarding low salinity water flooding, as an EOR technique, requires to view the injection which reflects positively on oil extraction though shifting chemical balance in the reservoir. This is instead of use water as just a ramp to move oil in a secondary oil recovery technique. While positive effects the low salinity water injection on oil recovery are known from 1940s the mechanism understanding and appreciation is relatively recent and the process widely recognized and researches in the last twenty years.

It should be stated that the practise shows that the reservoir response to injected low salinity water is very complex. Few processes with different names have been developed and used over the time: Low Salt (LoSal), Smart Water Flooding, Ion Tuning and Advances Ion Management. From the business point of view low salinity water injection speeds oil recovery by allowing oil to flow more easily through the rock and reduces amount of the injected water. As the reservoir development stage, the low salinity water can be injected after high salinity water flooding or even at the exhaustion of primary oil displacement drives. Number of applications in low salinity water injection in every of the last few years exceeds combined number of all method applications in the previous years all together.

Water salinity is an overarching term. Precise definition of it is not so simple and straightforward. In many cases the term is used as a synonym if the Totally Dissolved Salts (TDS) amount. Then it indicates how much (by weight) of various salts are dissolved in water but it does not say which kind of salts are dissolved.

Salt definition and behavior go far beyond everyday used term and properties referring to table salt or sodium chloride. In a broad definition, a salt is a chemical compound in which molecules contain cationic (positively charged) and anionic (negatively charged) parts. During dissolution in water salts can produce predominantly hydroxide ions (alkali salts), positive ions (acidic salts), or not change water pH balance (neutral salts). One can immediately see that salts by themselves are far from simple.

There are many salinity scales which are based on the easily applied measurement methods. The following information is provided just for a guidance. Some broadly accepted brackets for salinity are: rain water 20 mg/L; drinking water − below 500 mg/L; river water − below few g/L; sea water − 20 + g/L; connate water more than 60 g/L. One can immediately see that almost any water available at the ground level will have lower salinity than the connate (reservoir) water. One can immediately see, and this has been proven true in practise, that even see water can be regarded as low salinity compared to the most of connate water and can produce low salinity water effect to a point. Care should be exercises in assessment of divalent ions.

In many cases water salinity is measured by water electrical conductivity. Electrical conductivity measurements are used for well logging too. The reservoir (connate) water salinity increases with the well depth as a rule. Low water salinity from the deep reservoir will indicate the reservoir connection to the shallow water supply. It has been mentioned before that many chemical EOR implementations are sensitive to the reservoir water salinity. It should be stressed again that the connate water salt chemical composition and salt amount are very important and need to be determined before any EOR application.

The connate water mostly contains sodium, potassium, magnesium and calcium cations (Na^+, K^+, Mg^{2+}, Ca^{2+}). On the anion side it has chloride and sulfate radicals (Cl^- and SO_4^{2-}). Other anions and cations are presented too.

It is clear, that a low salinity water contains much-much less ions compared to a connate water. All surfaces and minerals containing water-soluble ions are affected by injected low salinity water. Low salinity water significantly affects reservoir clay. The effect is much clay specific as it depends on the clay type. The basic process is quite straightforward – many clays start to swell. Swelling by itself can make some pores impassable for liquids. Swelling in part also leads to delamination. Delaminated clay (so named fines from clay) becomes mobile and will block more pores. Overall effect will lead to development of very low or even zero permeability. Oil containing formations with significant clay content might be completely incompatible with low salinity water injection. Clay nature and clay content should be carefully analyzed before additional water injection into the reservoir.

At this point it is worth remembering that water interaction with the rock on significant part (some say on three quarters) defined by the interaction with ions on the surface (hydrogen bonds and dipole interactions with surface adsorbed ions). Rock wettability is determined to the great extent by the rock surface ions and their concentration. Low salinity water will change ion concentration and the surface ion speciation. It has been shown by many studies that most likely microscopic outcome of this change will be rock becoming more water wet. It is possibly immediately to guess that this leads to some oil detachment from the rock (rock now preferably coated with water) and will lower oil saturation after the sweep. It is reported that it is possibly to extract additional OOIP by the low salinity water injection. The reported extraction increase varies between 5% and 20% of OOIP in carbonate reservoirs.

The interest to low salinity water EOR for light to medium gravity oil is based on generally good results at low investment levels. This is assuming that the surface water is available at the required quantity and can be prepared for the injection. But even sea water has lower salinity compared to connate water. Addition of some chemicals to sea water makes the process even more efficient. The amount of injected water should be roughly just above 100% of pore volume. In many cases consecutive alteration during injection of sea and fresh water were reported to produce good outcomes.

14.2   Electrical thermal EOR

Electrical thermal Enhanced Oil Recovery is mostly based on supplying energy to the formation with the main goal to rise the temperature and reduce oil viscosity. It comes probably without saying than other than just pure thermal effect processes do take place, but they are difficult to evaluate and they are at the very beginning of the research activities.

The simplest electrical thermal EOR is direct reservoir heating with low frequency (50 or 60 Hz, country electrical grid standard dependant) electrical current when heat is produced by ubiquitous Ohmic heating. The process is very simple and straightforward. Connate water with high salt content conducts electrical current well and the water acts as a big heat emitting resistor. The only things needed are two electrodes to establish the current flow and big transformer to match the power. Two production wells can do the electrodes job. However current concentration near electrodes might produce overheating and in the overheated zones there will be water evaporation at first and then reduction of the process efficiency. To avoid this one electrode (or both electrodes) can be water injection or the connate water re-injection well. Horizontal well drilling technology allows installation of special heating cables precisely within the oil strata. The cables eliminate overheating problems and make the process very simple and economical.

This Ohmic method is very universal, its applicability is not limited by the reservoir geology, depth, pressure or temperature.

It is also possible to utilize inductive and microwave heating. In the first case, production well tubing is heated at the oil strata level by a medium frequency induction current. This allows to rise the temperature at the well adjoining zone and create flow zone with significantly improved oil flow near the production well. Low oil viscosity in this zone positively reflect on the general oil recovery. In the second case, microwave antennas can be places at the strategic part of reservoir to rise temperature in the zones with low permeability or reservoir temperature in general.

Electrical heating is usually consuming few times less energy than steam injection and has good potential to reduce the carbon footprint of the thermal EOR. Compared to the conventional steam injection electrical EOR can be implemented at any depth and does not depend on injectivity. It is also technically simple, less dependant on chemicals supply chain and does not produce waste materials.

14.3   Advanced polymer systems

Polymer flooding is well established and used EOR technique. The limitations come from the traditional polymer low robustness and the method sensitivity to the reservoir conditions. This usually requires to employ higher polymer concentrations to counterbalance the negative process sides. High polymer concentrations are more expensive by themselves, more challenging to inject down the well and create problems with the reservoir injectivity.

Many advanced polymer materials are constantly developed for various purposes. It is most useful for EOR to use polymers when the reservoir temperature instead of degrading polymeric solution would increase the solution viscosity, at least in some temperature range. In broad terms it is possible to talk about thermoviscosifying or thermo-thickening polymers. Additional benefit comes from the lower, as compared to traditional polymer, molecular weight to achieve the same viscosity. Additionally, some polymers from this class demonstrate low sensitivity to water salinity. In fact, some of thermo-thickening polymers show increase of viscosity at high salinity.

Other classes of new polymers will form nanoparticles at elevated temperatures, this in turn will increase the solution viscosity. It is also possible up to extend to use latent and delayed action polymerization systems.

All this allows, in principle, to manage highly permeable zones in stratified inhomogeneous reservoirs and to provide displacement in problematic areas. Injection of the polymer solutions close or into the challenging zones allows reservoir flexible management. The injection volumes then are not so big, and this allows relatively inexpensive additional oil recovery despite high advanced polymer prices.

14.4   Disperse systems

Creating some structures within or adding some particles to water modifies water flow (rheology) properties. The water flow properties start to deviate from Newtonian behavior. It is not exactly just viscosity increase. The system properties start to depend on shear rate, applied force and system history. For instance, viscosity can be decreasing over flow time (so named thixotropy), or the viscosity can increase over the flow time (rheopexy). The system behavior becomes very complex and can, in principle, be tuned to suit the purpose.

Disperse system behavior is defined by the dispersed particle interactions. If particles attract to each other than they will form some bigger conglomerates. At high enough concentrations this will lead to soft solid-like behavior. Repulsion forces between particles makes each particle to occupy certain, bigger than the particle, volume.

Many terms are used to describe disperse systems but they all have an unifying name − colloidal system. Thermodynamics and in many cases chemical reactions at the reservoir temperature define oil displacing properties of the systems with dispersed particles and structures. Earlier mentioned emulsions and foams also broadly fall into this category. Very important point to mention is that oil extraction is a dynamical process of matter movement through porous media. In general, porous media acts as a filter. Filter can either do nothing to the passing material (no interaction apart from hydrodynamics and capillary forces) or it can start accumulation part of the passing media. To add to the complexity the passing media (even oil from one part of reservoir passing through another part) can modify the filter (porous rock).

Properties of disperse systems are also very much defined by the concentration of dispersed media. Viscosity is be affected by the dispersed media concentration (volume fraction of dispersed particles) very dramatically at some concentrations. At low concentration (volume fraction of dispersed particles) the particles move freely by Brownian motion. The solution is then in a liquid state. The solution would have viscosity may be few times higher as compared to pure liquid. At around 50% concentrations particles can start forming crystalline domains. Liquid and crystalline phases co-exist. The viscosity rises up may be ten times. At around 58 vol fraction % the solution becomes a glassy solid with the viscosity rising by another hundred times or so. All mentioned properties only attainable in a static or at a very low shear rate. As shear rate increases the viscosity is reduced very significantly. As particle sizes get smaller the viscosity reduction on shear rate dependence is reduced. Particle shape and size distribution have complicated influence on viscosity.

In some systems there is an opposite effect of shear rate − the solution becomes more viscous with the increase in shear rate. Again, there are complicated dependencies on all parameters imaginable.

The particles not necessary to be a solid to create liquid with desired properties. The "granularities" can be another liquid droplets or even gas bubbles. It is possible to create carbon dioxide foam with some surfactant solutions. The foam has then high viscosity and when injected improves sweep efficiency of the reservoir.

We have briefly discussed emulsions before. They are just another dispersed system. Applying various methods and surfactants it is possible to create them for various purposes of oil displacement and extraction. Sometimes the emulsions are created in the production wells just by oil natural surfactants and then create problems for pumping.

Dispersed systems are very good advanced tool for EOR. The drawback is to get read of dispersed matter and surfactants after oil has been extracted. Even breaking water/oil emulsions is not exactly straightforward process. Interesting technologies start to appear with time-limited systems, when the dispersed system disappears by itself under reservoir conditions influence.

14.5   High pressure air injection (HPAI)

One active product is readily available for EOR − air. It is possible to inject air at a high pressure into an oil formation to achieve Enhanced Oil Recovery process. Oxygen from the air oxidizes oil and essentially produces flue gas and water. Flue gas then spreads through the reservoir mixing with oil and improving miscibility. High oil recovery by the HPAI is provided by many positive processes happening at the same time − the pressure in the reservoir rises, the temperature goes up, viscosity of oil drops, interfacial tension is reduced, oil swells due to carbon dioxide absorption, the displacement driven in part by nitrogen and water. The method is usually considered for light oil reservoirs as for heavy oil reservoirs in situ combustion is preferred. The rock formation significantly affects the method performance for many reasons. Also oil itself should preferably have low ignition temperature and have high oxygen utilization at low temperature oxidation. It is worth remembering that the oxidation process is faster at high pressure and this high pressure should be created and maintained in the reservoir. Metal catalyst particles also can be added to stimulate the oxidation process

The HPAI seems to be most attractive for recovery of tight oil. The method is not limited to depth and can be used for deep reservoirs.

14.6   Renewable energy EOR

All processes during oil extraction and during Enhanced Oil Extraction need energy. We have briefly outlined conventional energy sources for some EOR methods before. Renewable energy allows to reduce carbon footprint of all operations and has the potential to be practical economical choice. The renewable energy utilization is slowly incorporated in all oil extraction processes. The overall renewable energy price is constantly reduced and the price now became comparable with gas energy generation.

Currently, most energy for oil extraction is produced by electricity or by utilization of hydrocarbons. Solar light brings energy to many places at the amounts between 1 and 1.5 kW/m^2, this converts into an average of around 2000 kWh/m^2/year. It is feasible at the moment to generate electrical power from it with an economically viable efficiency between 10% and 20%. Electrical energy is very flexible in a sense that it can power many processes and machinery.

Much higher efficiency at solar light utilization is achieved by solar thermal technology. Nowadays some companies claim that around 90% of solar energy can be converted into the heat energy. This energy then can be used to heat a formation, for instance. In order to be an economically viable the site installation should be big enough and this requires significant financial outlay and considerable time to get the return.

Few projects in this area have been undertaken. Generally, the results are variable. Some installations have been successful while others were not so. As usual, the susses follows detailed planning, careful installation and site management. Many governments in oil producing countries have developed renewable energy deployment and use programs.

It is absolutely clear that the renewable energy use will be widened and successfully used in Enhanced Oil Extraction processes.

14.7   Heavy oil fields

The advantage of heavy oil is in its abundance. The disadvantage is in high viscosity and difficulties to recover it. Steam Assisted Gravity drainage (SAGD) has been developed and successfully used. In situ combustion is also very effective method, as was discussed before. Unfortunately, both of those methods are not good for thin formations. The oil community slowly realises that it is possible to use high viscosity displacement media such as surfactant-polymer flooding and high temperature dispersed systems for heavy oil recovery. An example can be a foamed steam which has much higher viscosity and allows much better sweep efficiency due to more stable oil displacement. Some surfactants are stable at temperatures above 200 °C. Added to a water with some noncondensable gas (the gas, nitrogen for example, partially stabilises bubbles) such surfactants allow to generate steam-foam and eliminate simple steam displacement front instability. It is possible to use other dispersed systems which properties can be further propped by addition of various catalysts.

14.8   Combined methods

Each EOR method by itself has its limitations. We have described before how combination of methods, like injection of various slugs, help to achieve fuller oil recovery. Some combinations have been used for long time and are well proven. There are numerous combinations which looks promising and are in the pilot studies stage. The potential is there in principle but adding combinations also require wide spectra of materials, equipment and, most importantly, expertize to implement them. The search for the optimal combinations and development of new equipment and methods no doubt will continue. The spectra of available techniques will inevitably allow to extract more hydrocarbons in all conditions and fields.

EOR modeling

15

Abstract

Simulation of the reservoir includes a description of the physical properties of reservoirs and fluids saturating them, consideration of the technological process for the productive layer exploration, creation of reservoir mathematical model and the actual computer simulation. All this allows one to predict the behavior of the reservoir and the oil extraction under various operating conditions.

Chapter Outline

Simulation of the reservoir includes a description of the physical properties of reservoirs and fluids saturating them, consideration of the technological process for the productive layer exploration, creation of reservoir mathematical model and the actual computer simulation. All this allows one to predict the behavior of the reservoir and the oil extraction under various operating conditions.

The need for reservoir modeling is justified by the need of oil companies for the most accurate prediction of reservoir development indicators under various operating conditions.

Reservoir modeling is important not only in order to save time and accurately forecast oil recovery, but also for a more in-depth understanding of the reservoir structure and behavior. This includes the flow-filtration parameters, the filtration properties of fluids and predicament of behavior ranges for these parameters during the reservoir development process.

In the classical approach, calculations are carried out on the basis of averaged parameters, since it is impossible to take into the account all dynamics of changes that ultimately affect development indicators. Using the capabilities of modern modeling programs and computing power, the field can be divided into millions of small cells. This allows to study the field behavior in great detail by applying filtering equations to each small cell.

Modeling of hydrocarbon fields mainly consists of two stages: static and dynamic modeling.

Primer on Enhanced Oil Recovery. DOI: https://doi.org/10.1016/B978-0-12-817632-0.00015-3

The static model includes the geological model of the field. The geological model is a framework of a field divided into millions of cells, where each cell contains all the petrophysical and reservoir-filtration parameters of a given field.

A dynamic model is a hydrodynamic model of a field. Data on a three-dimensional geological model, perforation, production, formation effects and other dynamic data are loaded into the hydrodynamic model. This allows for detailed forecasting based on the field development history.

15.1 Geological modeling

Currently, the main software packages for creating 3D geological models of oil and gas fields are: DecisionSpase and Geographix (Landmark); IRAPRMS (Roxar); Petrel (Schlumberger); Gocad (Paradigm) and few others.

The three-dimensional model is based on seismic survey data and the results of detailed correlation of well sections that are spatially interconnected (Figs. 15.1 and 15.2).

The traditional technology of 3D geological modeling includes the following main steps:

1. Collection, analysis and preparation of all necessary information, data download into the modeling software (import and export).
2. Strata correlation from the well drilling data.
3. Interpretation of seismic data (identification of inhomogeneity, horizon tracing and strata map building, attribute analysis, etc.).
4. Building and editing of full maps.
5. Building the tectonic disturbances model.

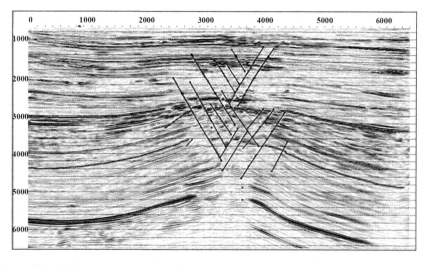

Figure 15.1 Oil formation seismic profile.

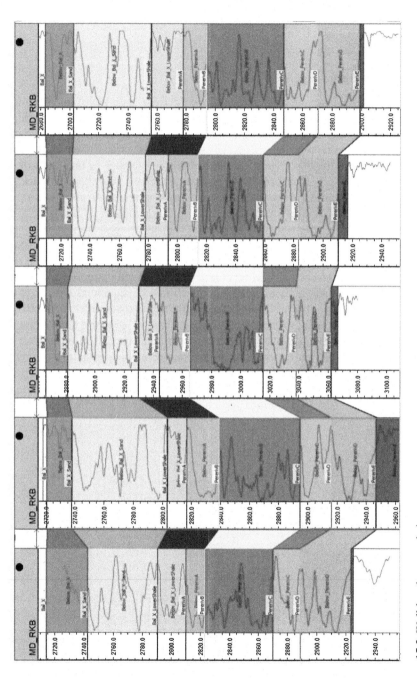

Figure 15.2 Well log data correlation.

6. Structural modeling (framing).
7. Creating a grid (3D grid), averaging (transferring) well data to the grid.
8. Facial (lithological) modeling.
9. Petrophysical modeling.
10. Calculation of hydrocarbon reserves.
11. Planning for production well placement and drilling.
12. Analysis of uncertainties and risks.

In order to create a three-dimensional model of the field, initial existing information is loaded (well coordinates, altitudes, inclinometry, well logs). After loading the source data and creating a working project, the structural-stratigraphic framework of the model is created (Fig. 15.3). For this, using the data of the reference seismic horizons as a trend, the wells are correlated beforehand (a breakdown of the formations in the wells is recorded). As a result, a tectonic disturbance model is created. The detailed correlation of the section is carried out on the basis of a series of transverse and longitudinal profiles covering the entire structure.

A structural model of the reservoir is created with all tectonic formations and discontinuations only after creating the structural-stratigraphic framework of the reservoir horizons. This allows to obtain structural maps of the roof and bottom of the reservoir and the external and internal contours of oil-bearing structure.

When studying sedimentation conditions, the principle of sequential stratigraphy is taken into the account, which allows to predict the strata units order. This study includes the sequence of the sedimentation process when the common chronostratigraphic system of process cycles is also taken into the account. The sedimentary complexes are determined on the basis of the internal reservoir geometry and their development in accordance with changes in the historic shelf environment.

Within the stratigraphic framework, on the basis of sedimentation patterns, a thin "slicing" of stratas is performed for each layer. This leads to creation of a

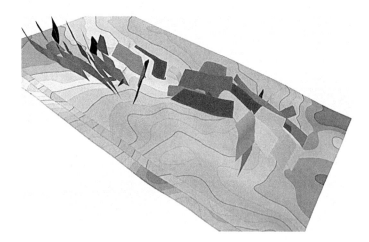

Figure 15.3 Structural stratigraphic framework of the horizon.

three-dimensional grid (3D grid). The grid is a cellular framework within which all the main stages of geological modeling take place.

The main difference between a three-dimensional grid and a two-dimensional grid is that each cell of a three-dimensional grid occupies a certain volume in space, whereas a cell of a two-dimensional grid is characterized only by the area. A well-constructed three-dimensional grid is the basis for building a correct geological model. The vertical dimensions of the model cells are selected taking into account the differentiation of the section according to filtration − capacitive properties. The vertical resolution of the grid is determined by the number of layers, which are selected in such a way that it was possible to most adequately restore the reservoir distribution space, without losing a single interlayer. The horizontal increments of the grid are chosen by taking into the account well placements, well density and the size of the whole oil field (formation).

Grid curves of facies, lithology, porosity, and oil saturation are transferred (averaged) to mesh cells along the drilled well trajectories.

As a result of modeling, based on the probability of occurrence of a particular conditions, each cell with the related parameters is assigned a code (number) of the corresponding (collector or non-collector) properties (Fig. 15.4). In addition, the most important part of indicator modeling is the definition of variograms for each property. Variogram analysis works better when a large number of wells is evaluated. At this stage, the modeling of reservoir properties of deposits is performed separately for each litotype of rocks (sandy-aleurite and clay, for instance), which makes it possible to identify clear boundaries when moving from one type of rock to another.

Then, a petrophysical model of the field is built (Fig. 15.5). The model is based on the results of the lithological modeling stage and allows to obtain consistent

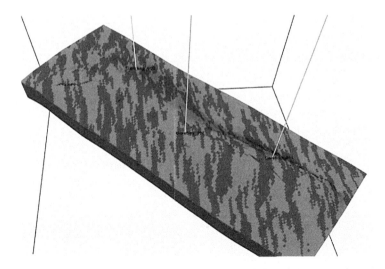

Figure 15.4 An example of lithographic properties of the horizon.

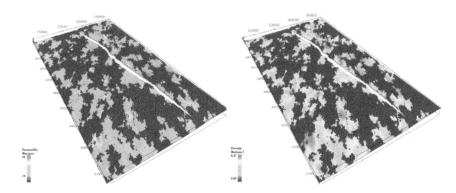

Figure 15.5 Petrophysical field model.

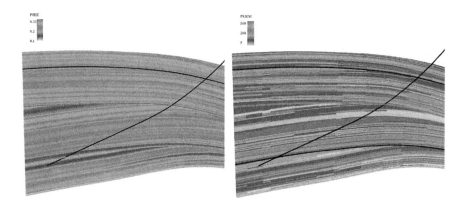

Figure 15.6 Reservoir properties distribution along well profile.

three-dimensional models of reservoir properties including oil and gas saturation in strata (Fig. 15.6). The model allows to see and analyze:

- correlation relationships between parameters (porosity, oil saturation)
- spatial heterogeneity of reservoir properties
- lithological heterogeneity of the reservoir formation
- geological patterns.

The model of the spatial distribution of reservoir properties (porosity, permeability, oil saturation) is one of the key components of a computerized 3D model of the oil field. The model is a key for correct estimation of initial geological reserves and for obtaining correct results in hydrodynamic modeling.

Also, the main result of the created 3D geological model is the ability to calculate the initial geological reserves of oil and gas.

Further, on the basis of the constructed geological model, a hydrodynamic model of the field is created, which allows flexible management and the development of an optimal oil field development strategy.

However, the main problem in the preparation of oil exploration project plans arises due to discrepancies between the static (geological) and filtration (hydrodynamic) models. It is clear that the models cannot produce absolutely identical results, newetheless, it is widely accepted, that the static and filtration models should differ from each other by no more than 3%.

15.2   Hydrodynamic modeling

The main software packages for creating hydrodynamic models of oil and gas fields are VIP and Nexus (Landmark), Tempest More (Roxar) and Eclipse (Schlumberger).

To create a hydrodynamic model of the horizon, first of all, the produced three-dimensional geological model of the horizon (oil strata geological model) needs to be loaded into the hydrodynamic simulator (Fig. 15.7). However, the multimillion cell "grid" built in the geological model is not feasibly acceptable for hydrodynamic calculations. The reason is that full calculations need to take into account too many variables and dependencies: like dependencies of the fluids physicochemical and thermodynamic properties on the reservoir temperature and pressure; like phase permeability variation functions. This calculation need to be done to each cell, and this leads to catastrophic growth of computation time. The solution to this problem is found in increment of cell size and reduction of cell numbers. This, in turn, will

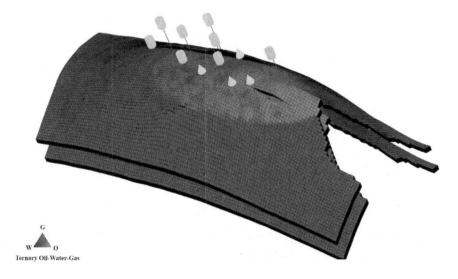

Figure 15.7 Saturation distribution for the hydrodynamical model.

lead to reduction of the calculation accuracy if addressed lightly. Balance between computation time and accuracy needs to be found for the each field case.

15.3   Hydrodynamic modeling for EOR

Further development (exploration) of deposits at a late stage of production, especially at the oil fields characterized by a complex structure and a significant proportion of residual reserves, is impossible without the use of Enhanced oil recovery methods. It is well understood and expected that the oil flow rate at the fields can be significantly increased due to modern methods of enhanced oil recovery.

Planning for the use of enhanced oil recovery methods is a complex task that requires a systematic approach to its solution. One of the important requirements in the planning of EOR before the start of the simulation are preliminary data analysis and the choice of proposed application method(s).

At present, the EOR method selection for development of an oil field is not a fully formalized procedure. At the same time, the choice of an effective technology for enhanced oil recovery for specific geological, physical and economic conditions of development is one of the most difficult tasks for a development engineer.

Before carrying out pilot works, it is advisable to have estimated data about the effect of the proposed implementation, to estimate by how much the oil recovery efficiency and other exploration parameters will change. Calculations and analysis using software are but essential steps in the development of oil and gas fields. In particular, hydrodynamic modeling is widely used to assess and predict the effectiveness of enhanced oil recovery.

To achieve this goal, mining and service companies conduct comprehensive hydrodynamic modeling in order to evaluate the effectiveness of proposed for the formation or strata enhanced oil recovery method(s). The following software packages are widely used: **ECLIPSE** and **INTERSECT** (Schlumberger); **Tempest MORE** (Emerson) **and Nexus**® **Suite** (Landmark).

ECLIPSE is used to create hydrodynamic models of oil and gas fields. The package allows you to calculate various development options, also to conduct the analysis of the effectiveness of the use of geological and technical measures and methods of enhanced oil recovery.

The ECLIPSE simulator supports a wide range of enhanced oil recovery methods, including thermal stimulation of a reservoir, chemical EOR, and also miscible and immiscible oil displacement by gas.

ECLIPSE modules have a wide range of functionalities, while they are easy to use and have a fairly simple interface.

INTERSECT EOR allows you to simulate high-resolution chemical EOR, which justifies its use in fields with various characteristics, such as the geology of the reservoir, the type of fluid and the strength of the aquifer. Numerous methods of EOR can be modeled using heavy oils or composite fluids for processes with low salinity, polymers and gas injection (including hydrocarbon gas, CO_2 and others).

TEMPEST is an effective and full-featured filtration modeling software with unique capabilities for creating multivariate models. It contains convenient 3D visualization, non-isothermal models, dual porosity/permeability models, compositional modeling, accounting for shale gas desorption, PVT data incorporation. The software can calculate well hydraulics, perform economic evaluation and much more. The program contains automated adaptations of geological and hydrodynamic models for the history of the development of an uncertainty analysis. Therefore, Tempest users cannot only create and adapt a single version (implementation) of the filtration model in the development history, but also obtain a representative multivariate field model, which can then be used to evaluate and optimize future production under conditions of uncertainty. Tempest also takes into account reservoir properties (heterogeneity, fracturing), filtration characteristics of reservoir fluids (non-Newtonian properties of high-viscosity oils, formation of a "plug" in the near zone of gas condensate wells), as well as the use of various methods of enhanced oil recovery (steam injection, water-gas effect, for example).

As it was said above, the application of any EOR technology at the field cannot be sensibly started with without prior analysis by the hydrodynamic simulators. The use of appropriate simulators is a necessary and very important step in the modern development of oil and gas fields.

EOR on site implementation

16

Abstract

All crude fields are unique. There are many EOR methods and the method choice should be made by taking into the account the reservoir conditions, available technical capabilities and production increment forecasts. There is always the best method to apply for all circumstances. Not all possible EOR methods have the same Technical Readiness Level (TRL). Big volume of knowledge and data should come together to make an informed choice. This chapter outlines steps which need to be made before full EOR project is implemented at an oil field.

Chapter Outline

16.1   Screening evaluation

The initial step before any Enhanced Oil Recovery method to a field (or more commonly to the specific reservoir or even oil containing strata) is to decide which oil production enhancement method or methods are most appropriate. This is done on the basis of reservoir and oil properties analysis — so named screening step. Screening studies are designed to assess the feasibility of using the enhanced oil recovery method based on a limited number of reservoir and fluid properties. Those properties are mostly regarded as critical. Screening is a comparison of the averaged characteristics of the reservoir with tabulated criteria of various methods applicability. The set of tabulated criteria is made on the basis of the joint international oil industry experience.

Most often, rather simplified technical screening is used to select an enhanced oil recovery method. In this case the analysis of geological and physical properties is conducted on the principle of "suitable − not suitable." In many cases this simple approach is most used as it is self-obvious and there is no need to use complex software. The boundary values are taken from the so named Taber tables. Taber and co-authors have in 1997 summarized the selection criteria for all major EOR techniques. They had analyzed the application results and tabulated the main applicability parameters. The authors identified 10 criteria values. Each criteria value is compared with the corresponding values for the field of interest. Those criteria are: fluid (oil) properties (API Gravity [^{0}API], viscosity [cp]) and reservoir related properties (oil saturation, oil strata thickness, location depth, temperature, pressure, porosity, permeability, lithography). Slightly modified assessment set, by recognizing the importance of the oil molecular composition for EOR application methods, can be summarized as presented in Table 16.1.

A score is assigned between 0 and 1 for each criteria. Grade 1 is assigned when the related property is within the recommended range and zero when it is not in this range. When a property matches only part of a range, the score is assigned in proportion to the matching rate.

Fig. 16.1 shows an example procedure for assigning scores. Using the deposit example we need to consider the main screening criteria for some conditional EOR. In this case all criteria with the exception of the permeability of the reservoir are within the specified ranges of screening and they are assigned the value 1. Permeability, as it is outside the recommended range, is assigned a proportional score equal to 0.6666. Next, we find the arithmetic mean value of the assigned points, which is 0.9524. Thus, the applicability of this method of EOR is 0.9524 points, i.e. quite high.

16.2   Phase behavior and core analysis

ToIn order to carry out full geological modeling, it is necessary to know the oil physicochemical properties, character of the reservoir, oil strata configurations and the reservoir rock filtration properties. For the complete understanding the geophysical studies should be complimented with laboratory analysis of reservoir fluids and core materials behavior at the reservoir conditions. All laboratory studies should be conducted at reservoir pressure and temperature.

16.2.1   Physicochemical properties of reservoir fluids

Complex reservoir fluids behavior is understood on the basis of laboratory experimental data and thermodynamic modeling. All this undertaking allows to predict fluid behavior in the feasible range of Pressure, Volume and Temperature (PVT coordinate space).

Reservoir characteristics

#	EOR method	Gravity (°API)	Viscosity (cp)	Composition	Oil saturation (% PV)	Formation type	Net thickness (ft)	Average permeability (md)	Depth (ft)	Temperature (°F)
Gas injection methods (miscible)										
1	Nitrogen and flue gas	>35 ↗ 48 ↗	<0.4 ↘ 0.2 ↘	High percent of C₁ to C₇	>40 ↗ 75 ↗	Sandstone or carbonate	Thin unless dipping	NC	>6000	NC
2	Hydrocarbon	>23 ↗ 41 ↗	<3 ↘ 0.5 ↗	High percent of C₂ to C₇	>30 ↗ 80 ↗	Sandstone or carbonate	Thin unless dipping	NC	>4000	NC
3	CO₂	>22 ↗ 36 ↗	<10 ↘ 1.5 ↘	High percent of C₅ to C₁₂	>20 ↗ 55 ↗	Sandstone or carbonate	Wide range	NC	>2500	NC
1–3	Immiscible gases	>12	<600	NC	>35 ↗ 70 ↗	NC	NC if dipping and/or good vertical permeability	NC	>1800	NC
Waterflooding (enhanced)										
4	Micellar/ polymer, ASP, and alkaline flooding	>20 ↗ 35 ↗	<35 ↘ 13 ↘	Light, intermediate, some organic acids for alkaline floods	>35 ↗ 53 ↗	Sandstone preferred	NC	>10 ↗ 450 ↗	>9000 ↘ 3250	>200 ↘ 80
5	Polymer flooding	>15	<150, >10	NC	>50 ↗ 80 ↗	Sandstone preferred	NC	>10 ↗ 800 ↗[a]	<9000	>200 ↘ 140
Thermal/mechanical										
6	Combustion	>10 ↗ 16 → ?	<5000 → 1200 → 4700	Some asphaltic components	>50 ↗ 72 ↗	High-porosity sand/ sandstone	>10	>50[b]	<11,500 ↘ 3500	>100 ↘ 135
7	Steam	>8 ↗ 13.5 → ?	<200,000	NC	>40 ↗ 66 ↗	High-porosity sand/ sandstone	>20	>200 ↗ 2540 ↗[c]	<4500 ↘ 1500	NC
—	Surface mining	7–11	Zero cold flow	NC	>8 wt% sand	Mineable tar sand	>10[d]	NC	>3:1 overburden to sand ratio	NC

$^{\ }$$c_1$...

NC = not critical.
Underlined values represent the approximate mean or average for current field projects.
[a] >3 md from some carbonate reservoirs if the intent is to sweep only the fracture system.
[b] Transmissibility >20 md-ft/cp.
[c] Transmissibility >50 md-ft/cp.
[d] See depth.

	Fluid and reservoir properties. Field under study (field a)	Screening criteria Taber-Seright (1997)	Screening results	
Viscosity, cp	500	< 100000	1	
API Gravity, °API	14	8–25	1	
Current Oil Saturation, fraction	0.6	> 0.4	1	
Thickness, ft	120	> 20	1	
Permeability, mD	150–300	> 200	0.6666	Score for each property
Porosity, fraction	0.38–0.52	NC	1	
Depth, ft	2900–3200	< 5000	1	
Pressure, Psia	600	NE		
Lithology	Sandstone (SS)	NE		
			0.9524	Score of the method

Figure 16.1 An example of simplified technical screening.

Properties of reservoir fluids are necessary for:

- oil and gas reserves estimation,
- well production rate forecast,
- field development design
- hydrodynamic field modeling.

The data allow to see at which PVT conditions the fluid can be a single phase or will separate into multiple components. This is especially important as simple equations, like Darcy's flow rate, only applicable to the single phase flow. Multiphase flow is described by much more complex than Darcy's equations.

The main parameters determined by PVT analysis for reservoir oil are:

- Oil saturation pressure with gas, methods of contact and differential degassing;
- Measurement of density of reservoir oil;
- Measurement of viscosity of reservoir oil;
- Paraffin crystallization temperature.

The main parameters determined by PVT analysis for a reservoir gas (gas condensate) are:

- The pressure of the onset of retrograde condensation;
- Measurement of the density of gas and gas condensate;
- Measurement of reservoir gas viscosity.

Laboratory analysis and thermodynamic modeling require specialized hardware, software and skillful personnel. There are many laboratories and companies specializing on properties analysis. It is probably possible to say that the consultancy analysis and equipment manufacturing for analysis field is dominated at the moment by three companies:

- Vinci Technologies
- Chandler Engineering
- Core Laboratories

16.2.2 Core analysis

To determine the permeability of reservoir liquids through formation rock at the reservoir conditions it is necessary to account carefully for several factors. These are: lithostatic or rock pressure, pore pressure and temperature. For creation of these conditions there are automated reservoir conditions modeling systems (equipment). The systems are designed and build for core research under conditions simulating reservoir. The analysis in general can be done in a wide range of specified values of pressures and temperatures.

All core studies can be divided into two large groups: standard and special core analysis.

Standard core analyzes are carried out in order to determine the parameters necessary for calculating geological reserves of oil and gas. This include oil and water saturation, porosity, permeability, pore size distribution, saturation distribution pattern over OWC (oil-water contact angle values).

Special core studies are carried out in order to obtain reliable baseline data for hydrodynamic modeling of deposits: relative phase permeability during two- and three-phase filtration; oil displacement factors with water, accounting for driving/ dissolved gas and various injected chemicals.

The main manufacturers of core analysis systems are:

- Vinci Technologies
- Coretest Systems
- Core Laboratories

16.3 EOR implementation

After conducting a hydrodynamic simulation and obtaining positive results for the proposed EOR in the pilot area, the project proceeds to the full implementation.

Usually, a pilot site with an injection well (typically a 5−7 point system is used) with surrounding production wells (see Fig. 16.2) is taken at the field.

In order to monitor and evaluate the process and determine the mechanism before and after the start of the event, a number of necessary studies are carried out at the wells. Before the start of the event, production logging studies are carried out at the production well to assess the vertical sweep efficiency.

The process influence efficiency should be assessed by collecting and comparing data on all processes. Usually Production Logging (so named PLT, where T can stand for Technique or Tool) is implemented and employed. With certain periodicity PLT logging studies need to be repeated at all well (exploration, injection, production) (see a log examples at Fig. 16.3).

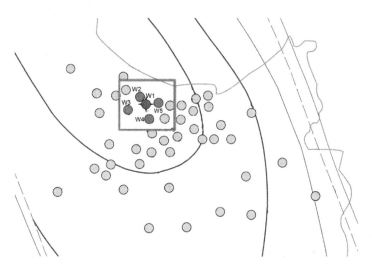

Figure 16.2 Pilot implementation at an exploring oil field.

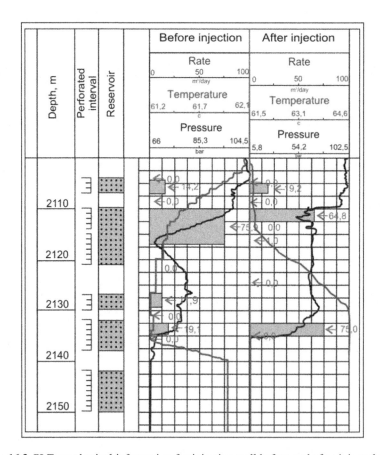

Figure 16.3 PLT geophysical information for injection well before and after injected slug.

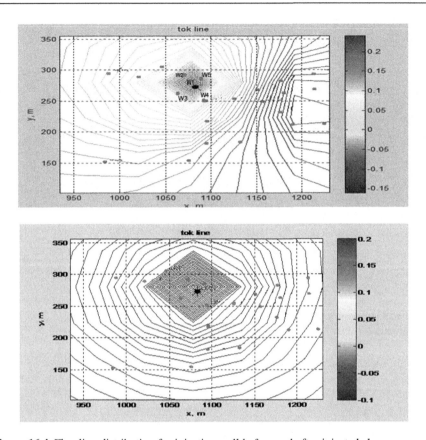

Figure 16.4 Flux line distribution for injection well before and after injected slug.

To estimate the influenced area (areal sweep efficiency), graphs of the distribution of streamlines (fluxes) are constructed (see Fig. 16.4). As can be seen from the obtained curves, before the impact, the injected water had moved in one direction (a) and after the injection event the flow lines have a uniform distribution over the area (b).

To determine the additional oil production, the dynamics of the flow rate of oil and water are plotted before and after slug injections (see Fig. 16.5). The additional material production from the wells is closely monitored (see Fig. 16.6)

If the pilot plant production increases against "business as it was" (projected decline curve) at the magnitude above 10% then the applied method can be extended on the whole formation.

16.4   Technology readiness level (TRL)

Concept of Technology Readiness Levels (TRL) was first developed in the USA by NASA. Now it is widely taken as the basis philosophy for the new technology

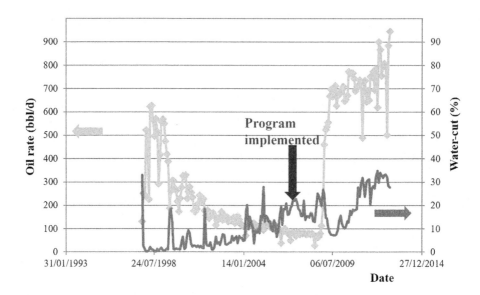

Figure 16.5 Dynamics of oil production and water-cut at the injection program implementation.

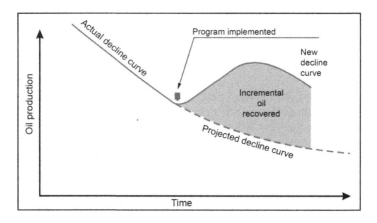

Figure 16.6 Assessment of incremental oil produced.

graduation development stages by many agencies and organizations worldwide. The classification reflects the state of development results for wide production and/or application. Assessment of TRL and current TRL assignment indicates to the markets and consumer readiness of the process/technology for wide market implementation. Knowledge of TRL makes

Table 16.2 Technology readiness levels for EOR technologies.

Phase	TRL	
R & D	TRL 1–3	Colloidal dispersion gels (CDG) injection Performed particle gels (PPG) injection In situ upgrading (heating/catalytic) Delayed action polymers (DAP) injection Hybrid processes
Demonstration–optimization	TRL 4–8	High pressure air injection (HPAI) Thermo-active polymers injection Low salinity waterflooding Water alternate gas injection Microbial Polymer disperse systems (PDS) Alkaline/surfactant/polymer SAGD
Well established EOR processes	TRL 9	Polymer flooding Thermal simulation Miscible gas processes Steam injection

it easier for developers and customers to monitor the progress of research and the choice of technologies that are most ready for industrial application.

There are nine levels of technology availability. Levels 1 through 6 are technology developments that are carried out as part of research projects. From the seventh level and higher, industrial development work begins, or a demonstration of the performance of technologies on real developed devices.

For the petroleum industry, the TRL approach to assess the maturity of software technology was first applied by TOTAL.

In this approach, the maturity of technology has 9 levels too.

Levels 1–3 represent the "Research" stage (Research: Basic principles, Concept and application formulation, Concept validation), 4–6 represent "Development" (Development: Experimental pilot, Demonstration pilot, Industrial pilot), and grades 7–9 – "Deployment" (Deployment: First implementation, A few records of implementation, Extensive implementation).

Many oil companies use this TRL methodology to assess the maturity of oil and gas production technologies.

Table 16.2 shows the level of TRL for various enhanced oil recovery technologies.

Before committing to an EOR project, all available EOR technologies need to be carefully compared in view of the particular reservoir properties. Business considerations need to be taken into the account too. The Table 16.3 provides but an example of a structured decision undertaking process.

Table 16.3 Assessment of eor technology for implementation.

Company name				
			Polymer flooding	
Date	8	Theme SPA/s	Reservoir performance	
Current TRL				
Technology description				
Relevant experience				
Company position				
References			Potential deliverables	Timeline
Value proposition				
Production increase				
Reserve growth			Risks/issues	Mitigation/contingencies
CAPEX reduction				
OPEX reduction				
Integrity and risk mitigation				

Further reading

Al Arfi, S.A., Mohamed, O.Y., Keshka, A.A., Salem, S.E., Al-Bakr, S., Asmar, M.S.E., Al-Amiri, A.A. Inflow control device an innovative completion solution from "Extended Wellbore to Extended Well Life Cycle. In: Paper SPE-119599-MS Presented at the SPE Middle East Oil and Gas Show and Conference, Manama, Bahrain, 15−18 March 2009.

Irap RMS 2009.1. User Guide. Stavanger, Norway, 2009.

Kjeilen-Eilertsen, G., Merzi, T., Burgos, M., Brönner, U. TRL (technology readiness level) assessment of DREAM (dose-related risk and effects assessment model) to qualify its use for modelling of produced water and drilling discharges. In: SPE International Conference and Exhibition on Health, Safety, Security, Environment, and Social Responsibility, Stavanger, Norway, 11−13 April 2016.

Romero-Zerón, L., 2012. Advances in enhanced oil recovery processes. In: Romero-Zerón, L. (Ed.) Introduction to Enhanced Oil Recovery (EOR) Processes and Bioremediation of Oil-Contaminated Sites. ISBN: 978-953-51−0629-6.

Shandrygin, A.N., Lutfullin, A. Current status of enhanced recovery' techniques in the fields of Russia. In: SPE Paper 115712 presented at the SPE Annual Technical Conference and Exhibition. Denver, Colorado, 21−24 September 2008, 18 p.

Shokir, E.M., Sayyoch, M.H. Selection and evaluation EOR method using artificial intellegence. In: SPE Paper 79163 Presented at the Annual International Conference and Exhibition. Abuja, Nigeria, 5−7 August 2002, 7p.

Suleimanov, B.A., Latifov, Y.A., Ibrahimov, Kh.M., Guseinova, N.I., 2017. Field testing results of enhanced oil recovery technologies using thermoactive polymer compositions. SOCAR Proc. (3), 17−31.

Taber, J.J., Martin, F.D., Seright, R.S., 1997. EOR screening criteria revisited-part 1: introduction to screening criteria and enhanced recovery field projects−part 2: applications and impact of oil prices. SPE Reservoir Eng. pp. 189−198; 199−205.

Zakirov, E.S. Upscaling in 3D computer simulation. M.: ZAO "Books and Business", 2007.

Zakrevsky, K.E., 2011. Geological 3D Modelling. EAGE Publications, Netherlands, ISBN: 978-90-73781-96-2.

<http://www.vinci-technologies.com/>.

<https://www.chandlereng.com/>.

<https://www.corelab.com/>.

<http://www.coretest.com/>.

EOR economics

17

Abstract

Contemporary oil recovery faces with three choices — to find new oil fields, to improve recovery factors from the well-developed fields and to progress in recovery from heavy oil fields. It proves increasingly difficult to find new easily (e.g. cheaply) recoverable oil. It is estimated that only below one third of all existing oil can be counted as light and is relatively easy to recover. More and more attention is diverted to further development of already used fields or to develop heavy oil reservoirs.

Contemporary oil recovery faces with three choices — to find new oil fields, to improve recovery factors from the well-developed fields and to progress in recovery from heavy oil fields. It proves increasingly difficult to find new easily (e.g. cheaply) recoverable oil. It is estimated that only below one third of all existing oil can be counted as light and is relatively easy to recover. More and more attention is diverted to further development of already used fields or to develop heavy oil reservoirs.

Oil recovery is well established activity in majority of technical and financial aspects. Enhanced Oil recovery is but a stage of oil recovery in general. Implementation and exploitation of EOR are associated with an additional field development and financial outlay. It can be viewed as an integral or a separate stage of oil recovery from a field.

During EOR implementation the old injection wells might need to be converted for the new types of injection. New wells might be required. Solution preparation, additional liquid/gas distribution network, new injection equipment and monitoring infrastructure will be needed. Supply chains need to be established and necessary chemicals purchased and supplied. Produced oil might need additional separation with bigger capacity and special treatments. Higher volumes of connate water should be handled and, probably, re-injected. The later by itself might need more injection wells and pumping capacity. All separated from oil materials need to be carefully assessed and re-used if possible. At the later stages of the field exploration a recycling of the used equipment and infrastructure will be required to satisfy the environmental protection. Many activities will necessitate more site personnel. The site operation and maintenance will be more complex and require more management and additional financial outlay. This is not the full list of everything what is needed but it paints the seen.

The financial package is relatively complex and includes many components. It is strongly advisable to gather as much information as attainable about the reservoir,

Primer on Enhanced Oil Recovery. DOI: https://doi.org/10.1016/B978-0-12-817632-0.00017-7

fluids, the reservoir conditions and make many laboratory tests. Feasibility studies need to be undertaken. Similarity data (data on application for the very close conditions), if accessible, are one of the best implementation guidance. The gathered information will make a solid foundation for the application and will allow to reduce the risk premium.

Challenges are both, technical and financial. What is completely clear that the Enhanced Oil Recovery from the formation will add to the price of barrel of oil and might reduce the profit margin, at least on the short run. There are too many price and tax variables in the overall financial model to make the general statements of encompassing long term profitability. Many EOR processes require time to develop and one needs to project production, investment and financial returns for a quite few years ahead.

Much information on costs were published by the companies on EOR results. The amounts are spread roughly from 10 to 40 US$ of additional cost per barrel. The EOR project is big investment and it requires decent financial modeling. It is prudent to plan for failure at the initial stages of implementation. It is much better if the initial plan accounts for possible "not as good as expected" results and has clear strategies to return to the planned oil production.

Environmental concerns ought to be addressed by all industries, the oil production and further processing are no exception. It is impossible to make outright judgement on the overall impact of new technologies as all commodities, equipment and taxation policies change constantly. It is clear that the oil extraction will continue for the foreseeable future. New technology and all processes will re-balance and will eventually find an average value which will reflect in the price of crude.

Crude oil is probably the most volatile commodity. The oil price and production costs vary nationally and internationally. In many cases taxation regime makes from big to huge difference on the overall project returns. This is especially important for the Enhanced recovery. Environmental concerns will make significant impact on the whole oil business and this in the much greater extend will affect Enhanced Oil Recovery and its place.

Index

Note: Page numbers followed by "*f*" and "*t*" refer to figures and tables, respectively.

Printed in the United States
By Bookmasters